CW00455712

King Edward Mine

by

Tony Brooks & John Watton

An Illustrated Account of
Underground and Surface Operations
1897-2001

This book is dedicated to the memory of William Thomas and John Charles Burrow, without whose efforts and friendship over 100 years ago this book could not have been written.

King Edward Mine

First Published in 2002 by
Cornish Hillside Publications
St Austell, Cornwall.

ISBN 1 900147 27 0 (clothbound)
ISBN 1 900147 26 2 (paperback)

Design and cover by
The Design Field (www.thedesignfield.com)

Printed by Short Run Press Ltd., Exeter, Devon.

Contents

Foreword

King Edward Mine was never a large mine: by present day standards it is almost insignificant in size; but its influence on the world of mining has been immense. Indeed, much of the world's mining has later been done by mining students who were fortunate enough to have trained there.

The authors have laid out a grand survey of why the mine came into being as part of the Camborne School of Metalliferous Mining, later just the Camborne School of Mines, its evolution and physical structure. The display of photographs that they have put together bring the text to life. In the text we are led through the background details regarding the mine and its plant: it also follows the history of a number of past students: a most revealing set of biographies.

The account is built on stages representing the development of the mine, and how teaching took place. This is entirely appropriate as one can see, in retrospect, that these represent the changes which were necessary to meet the needs with the evolution of mining techniques and conditions. I experienced the "Marking Time" era, in the early 1950's. Others from previous and later phases no doubt had different experiences from my time. The 'atmosphere' was possibly not that which present educational theorists would approve of – but certainly had much to be grateful for that it was as it was. There was a remarkable tolerance both by staff and fellow students towards a very wide variety of 'types'. The student body was completely international. Thus one, who was usually less substantial than a shadow; another, whose huge South African frame ploughed all before it on the rugger field. Both were nurtured by the Mine and School. It is worth noting of these two that one became of major importance as a consultant on mining matters to a large Commonwealth country, the other the General Manager of some of the biggest and deepest gold mines in the world. Whatever theoretical educationalists might like to say, something certainly was done right.

The mine staff were carefully selected. Miners from the four corners of the earth enjoyed passing on all sorts of knowledge to apprehensive students: how to hand drill, charge and fire a shothole to what crooked tricks could be practised on or by miners – valuable information gleaned during work underground or over a cup of tea after the 'dry'.

The account of mine photography is fascinating. In the early days of the mine photography was used as a source of illustrations that were valuable teaching aids. John Watton is the present day bearer of, almost literally, the 'torch' passed on by the great Burrow and his friends. One has only to try and take photographs underground to appreciate the skills required.

From the gold ring on a lady's finger to the Channel Tunnel, via various wealth creating affairs such as work on discovering North Sea Oil, we all owe much to King Edward Mine.

Bryan Earl
January 2002

Introduction

During the early 1890's students at the Camborne School of Mines (then known as the Camborne Mining, Science and Art School) did much of their practical work in the local tin mines. South Condurrow Mine, being close to Camborne and quite shallow, was extensively used in this way.

In 1896 a case was made for the School to have its own underground mine. South Condurrow was about finished as a commercial concern and, in 1897, the School took over the abandoned eastern portion of the mine. In the period between 1897 and 1906, under the direction of William Thomas, Head of Mining and Surveying, new buildings were erected on surface and the underground section of the mine around Engine Shaft was reopened and brought into production.

In the early 1890's John Charles Burrow, a local photographer who had premises in Camborne, was developing his skills in mine photography with special emphasis on underground scenes. William Thomas, no mean photographer himself, was a friend of Burrow and he used to accompany Burrow and another photographer friend, Herbert Hughes on their excursions to mines around the County. William Thomas is seen here in **Plate 1** with in the centre J.C. Burrow and to the right, Herbert Hughes. Thomas and Burrow collaborated in the splendid book of underground scenes, 'Mongst Mines and Miners', published in 1893, where Thomas wrote the text and Burrow took the photographs. Thomas no doubt encouraged Burrow to photograph 'his mine'. Between 1893 and about 1905 Burrow took a whole series of photographs of the mine both on surface and underground. As a result of this collaboration between Thomas and Burrow, and to a lesser extent Hughes, King Edward Mine is one of the most photographed mines in Cornwall. Many of these photographs are used in this book.

Plate 1. L-R W Thomas, J C Burrow and H Hughes

Relatively few photographs of the mine exist for the period after 1906. It cannot be coincidence that in 1906 William Thomas left the Mining School to take over the management of the Botallack Mine. Between 1907 and 1908 Burrow took a number of photographs of the developments at Botallack. These photographs have also survived. Thomas left Botallack in 1908 and the series of photographs ends at the same time.

The authors of this book feel privileged to have followed in the footsteps of Burrow and Thomas.

John Watton was, for a number of years, Head of Audio Visual Aids at Camborne School of Mines and he, like Burrow, was awarded his Fellowship of the Royal Photographic Society for work on underground photography. It was from John Watton's original suggestion, in 1987, that the present scheme to preserve King Edward mine as a museum was developed. Tony Brooks, a mining engineer was, like Thomas, Head of Mining & Surveying at Camborne and also was the School's Mine Manager. He worked with John Watton on some of his underground photography and is a founder Trustee of the Botallack Trust which has carried out preservation work at Botallack – William Thomas's old mine.

This book primarily tells the story of King Edward Mine over the past 100 years. Also included are the underground workings at the Great Condurrow Mine and the Holman Test Mine. These mines have been used by the School of Mines since the underground section at King Edward was lost in 1921. The objective has been to weave the photographs into a history of the mine using the photographic images as the basis for the explanation of what, how and why things were done. We have also taken the opportunity to examine the early history of, and techniques used in, underground photography in Cornwall and to introduce some of the photographers who worked at King Edward.

Acknowledgements

This book could not have been written without the help, encouragement and enthusiasm of many people including:

Jay Foote and the ever helpful staff of the CSM library, who have traced the most obscure references and presented them with a smile.

Terry Knight and Neil Williams at the Cornish Studies Library – which is now finally housed in much deserved new premises at Alma Place, Redruth.

David Thomas (archivist) at the County Records Office for assistance with image research.

Rob Cook (curator of the photographic collection) at the Royal Institution of Cornwall for image research and making image files available.

Gill Thompson (librarian) at the Royal Photographic Society for research and copying documents and articles by J.C.Burrow FRPS and H.W.Hughes FRPS.

Justin Brooke who, as always, gave us full access to his extensive files.

Tom Morrison for permission to use material from his book 'Cornwall's Central Mines. The Southern District 1810-1895',. The Camborne School of Mines Association for permission to quote from 'A Student in the Klondike and South Africa'

Permission to publish photographs from: Paddy Bradley, Camborne School of Mines, Tony Clarke, Cornish Studies Library, Cornwall County Records Office, Bryan Earl, Royal Institution of Cornwall, John Rapson and Courtenay Smale.

Because the photographs are such an important aspect of this book we have tried, where possible, to establish the names of the photographers, the location of original negatives, the location of prints from the original negatives or failing that the location of the prints that we have used. These are listed in Appendix III

Much of the detail of the early history comes from the Camborne School of Mines Archive. This includes minutes of various committees of the Camborne Mining Science and Art School and the School of Metalliferous Mining and the School of Mines' Magazine and the School's prospectuses which can be found in the School's library. Additional material came from the pages of the Mining Journal and the Mining Magazine. Roger Penhallurick prepared the location map used as Figure 2.

Finally to our publisher, Charles Thurlow, a past Camborne student and mining engineer, whose knowledge of the subject and attention to detail has contributed so much to this book.

References

Anon. A Student in the Klondike and South Africa. Camborne School of Mines Association Journal, 1997-98, p11-13.

Beresford Lees, P., On the Geological History of William's Lode in King Edward Mine, Cornwall. Trans. Royal Geological Society of Cornwall, Vol. XIII, 1914.

Börner, H., Der Bergmann in Seinem Berufe. Freiberg, 1891.

Burrow, J.C. & Thomas W., 'Mongst Mine and Miners. Simpkin, Marshall, Hamilton, Kent & Co., 1893.

Clarke, A.J. & Rabjohns, E.W., The History of Wheal Grenville. CSM Journal Vol.74, 1974.

Howes, C., To Photograph Darkness. Gloucester, 1989.

Hughes, H.W. Textbook on Coal Mining. London, 1904.

Hughes, H.W., Photography in Coal Mines. The Photographic Journal, 25 November, 1893.

Newhall, B., The History of Photography. New York, 1982.

Morrison, T.A., Cornwall's Central Mines. The Southern District 1810-1895. Alison Hodge, 1983.

Piper, L.P.S., A short history of the Camborne School of Mines. Trevithick Society Journal No.2, 1974.

Thomas, H., Cornish Mining Interviews. Camborne, 1896.

Watton, W.J., History of underground Photography in Cornwall. Journal of the Trevithick Society, 1990

List of Figures

List of Plates

Chapter 1
The Search for a Mine

The Camborne School of Mines
The Camborne School of Mines (CSM) evolved from attempts in the 1850's to improve the technical education of miners in Cornwall. The Miners' Association of Cornwall and Devon was formed and classes held in a number of centres. In 1863, the Association adopted the National Science and Art Department schemes so that students could obtain recognised certificates. The scheme can be described as a 19th Century technical college system. In 1882, as the awareness of the importance of technical instruction grew, work was started on building the Camborne Science and Art School (**Figure 1**). In 1888 the Camborne Mining School was established here as well. Mining education was also carried on at the Science and Art Schools in Redruth and in Penzance.

The Principal of the Camborne Mining School was J.J.Beringer, whose special subject was assaying, which figured largely in mining education of that era. William Thomas was in charge of mining, ore dressing and surveying. William Thomas was the son of Charles Thomas who had been the manager of the Cooks Kitchen Mine near Camborne. 'Willie' Thomas, who had gained early experience working with his father, was described in the 1881 census as a 'Land & Mine Surveyor & Clerk Tin Mine'. He was something of a character, who had strong opinions as to the way things should be done and did not always agree with the managing authorities. **Plate 2** shows what is thought to be the staff of the School of Mines around 1900. William Thomas is sitting in the front row grasping an umbrella and immediately to his left, wearing a soft hat is J.J. Beringer. At the far left of the photograph, perhaps looking a little uncomfortable, is Capt. Nicholas Temby, the Mine Foreman.

Figure 1. School of Mines letterhead

Plate 2. Staff of the Mining Science and Art School about 1903. Centre wearing the soft hat is JJ Berringer, head of the Mining School and to his right with the beard is William Thomas. Far left is Nicholas Temby the Mine Foreman

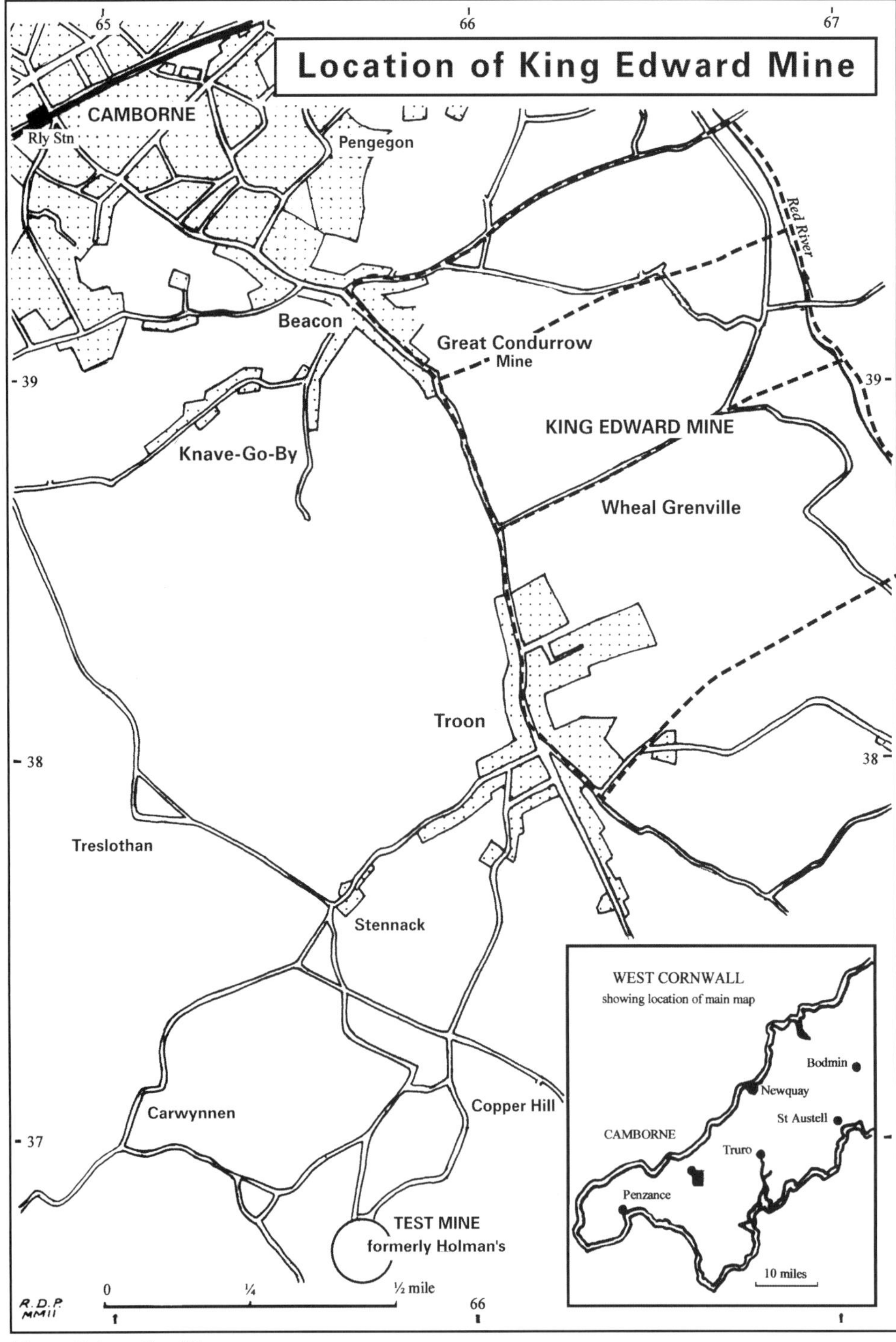

Figure 2. Location Map

In the early stages of the School's development the greater part of the theoretical and lecture work was performed during the evenings, while the practical side of the training was undertaken during the daytime, the students being placed in different mines in the locality. In the early part of the 19th Century, Cornwall was a major copper producer and led the world in deep hard rock mining, mine drainage and engineering. By the end of the century the copper industry had collapsed and tin mining, which had expanded greatly after 1850, was in serious decline. The surviving mines were under capitalised and were falling behind technically. This environment was hardly ideal from the instruction point of view, and it was difficult to supervise the training that the students were getting. This was particularly so for the technical subject of underground surveying. It was realised that the only real solution was for the School to have its own underground mine. On the 2nd of July 1895, Messrs Beringer and Thomas were asked to formulate a scheme to overcome this problem.

A number of alternatives were examined in the search for a site which offered readily accessible shallow workings with little or no drainage costs. Eventually, in 1897, an agreement was reached with the South Condurrow

Company to take over the eastern portion of their mine near Troon, which had been abandoned some years previously (**Figure 2**). South Condurrow after 30 years of profitable operation was almost worked out and had suspended operations in May 1895 due to the collapse of the tin price.

South Condurrow was in many respects ideal. It was close to Camborne and the land and mineral owner was Mr. W.C. Pendarves, who was closely involved in the development of the School. Students had for some years carried out surveying and practical work underground at the mine. The mainstay of South Condurrow had been the Great Flat Lode which outcrops in the northern part of the sett and, dipping south at 30 to 35 degrees to the horizontal, passes into Wheal Grenville in depth to the south. The greatest advantage was the fact that the mine did not need to be pumped. In the 1880's Wheal Grenville had mined across the boundary into South Condurrow and had holed into the lower workings so that all water from the eastern part of the mine gravitated to the Grenville pumps.

History of the South Condurrow Mine to 1896

South Condurrow lies on the northern slope of a shallow valley that runs eastwards down to the Red River. From just north of the village of Troon a straight road runs about 800 yds. east-north-east down the centre of this valley and marks the boundary between South Condurrow to the north and Wheal Grenville to the south. The sett is about 300 yds. wide and extends from just west of the Camborne-Troon road eastwards to the Red River.

Of the vertical lodes, only two were of significance namely, William's Lode and King's Lode. William's Lode was worked from Engine Shaft, vertical to 40 fathoms below adit and William's Shaft which was vertical to adit and then followed the lode down to the 20 fathom level. King's Lode, to the south of William's Lode, was worked from Plantation Shaft, Vivian's Shaft, King's Shaft about 100 yds south of Engine Shaft, Tye Shaft, Frazer's Shaft, and New Shaft. (**Figure 3**)

Whilst the whole area is underlain by granite, the immediate rock below surface is killas or clay slate. The killas, nowhere more than 400 ft. deep, lies in a trough between two granite ridges – on the north forming Beacon Hill and on the south leading into the main mass of the Carn Menellis granite.

The first recorded production from South Condurrow was in 1864. Copper ore sales were always small and the mine only produced some 1050 tons in the period 1864-1882. The first recorded production of black tin was in 1865 and sales did not become substantial until 1870. Initially, the mine had no tin concentration plant of its own and an eight-head water powered stamp battery was rented. At this time the mine was quite small with 39 men and 6 boys underground and 20 people on surface.

The first main shaft, started in 1864, was Engine Shaft which was sunk to a final depth of 40 fathoms below adit or 400 ft. from surface and was equipped with a 45-ins pumping engine. Vivian's Shaft, in the centre of the mine, was sunk at the same time.

All the shafts at South Condurrow experienced serious ground stability problems with frequent collapses in the shafts caused by the weak waterlogged clay-slate killas and granite near surface. King's Shaft, 300 ft. south of Engine Shaft and named after the company's secretary, was started over an extension of the West Basset Lode. This was vertical to adit and then followed the lode down dip, i.e. in the lode.

Shafts sunk following the lode were very common in Cornwall in the 19th Century. In the short term they were the cheapest option in the hand-to-mouth way that mines were generally operated at that time. If the lode looked 'kindly' then the decision would be made to deepen the mine. Many are the agents' reports, aimed at anxious investors, recording good values of lodes in shafts being sunk. Whilst they provided instant prospecting they created serious pumping and winding problems for the future. Lodes are not linear structures with a constant dip. The dip may steepen or flatten out or even be displaced by faults. Pumping, before the days of electricity, required the installation of a reciprocating pump rod down to the bottom of the shaft in order to transmit power, from the beam engine on surface, down to the pumps placed at regular intervals down the shaft. Crooked shafts caused increased friction losses in what was already an inefficient system. Similarly fast winding of ore in skips was impossible.

King's Shaft was equipped with pumps apparently operated by flat rods from the pumping engine on Engine Shaft. Early in 1868 it is recorded that a steam winding engine, already on the mine, was removed to a new site. This almost certainly was to Engine Shaft and the 'new' engine house is the one that still stands on the site today. It is sited between Engine and King's shafts. It is probable that this winder was used to wind from either shaft as required.

The story of South Condurrow is of a never ending battle against water. For many years the mine depended upon the pumping engine at Engine Shaft

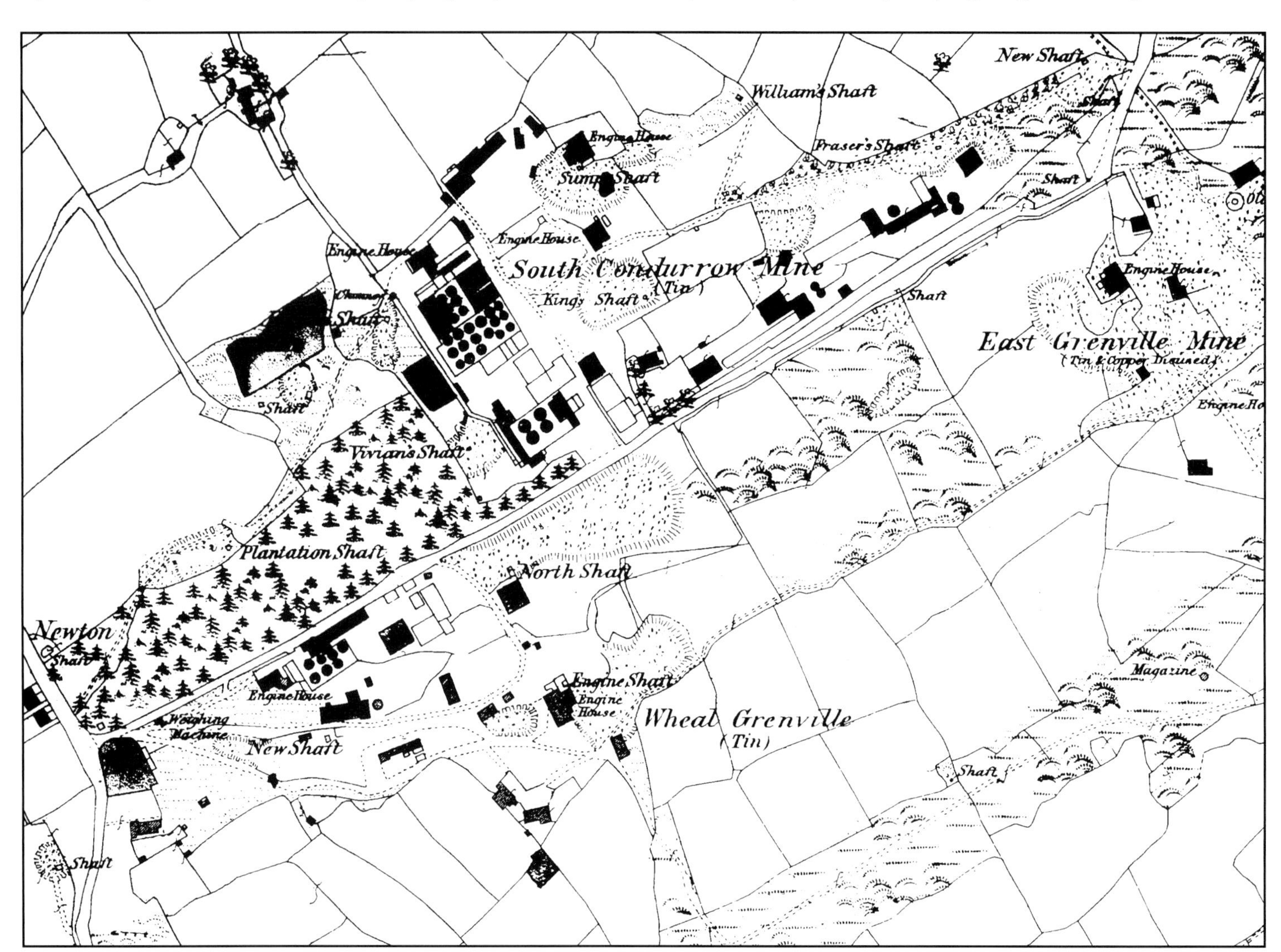

Figure 3. 1880 O.S. Map of Mine Area

to operate pumps in several shafts by means of flat rods and wire ropes. Despite these problems production continued to increase and it was decided to erect steam powered stamps on the mine and to cease to lease the stamps on the river. Forty-eight stamps were purchased and the engine and stamps were fully operational by November 1869. Dressing floors and a calciner were also built. This stamping engine was to serve the mine until the end of its life.

Tin dressing requires a lot of water and South Condurrow was typical of Cornish mines of the period – too much water underground and not enough on surface. Partly to overcome this shortage on surface, a small shaft was sunk immediately in front of the stamps engine house. This shaft connected with deep adit and, using the same engine, water was pumped up to the stamps and dressing floors. This shaft and the consolidated ruins of the stamps engine house can be found a short distance west of the present King Edward Mine buildings.

Tin production was now beginning to increase rapidly as the Great Flat Lode was being exploited, and the sales of black tin increased from 51 tons in 1869 to 219 tins in 1870. By now King's Shaft was the main production shaft for the eastern and central part of the mine.

The shallow dip of the Flat Lode took it south through the boundary into Wheal Grenville at around the 90-fm. level. At some point the two mines holed into each other, Grenville being seen as the worse offender. This would have presented little problem at the time with South Condurrow wanting to exploit its steep lodes below the 90. However, if South Condurrow were to stop pumping then the water would rise to the 90 and overflow down into the Grenville workings.

By 1874 some 18,000 tons per year or 50 tons per day was being mined with 110 men underground. This can be compared with the modern South Crofty mine which milled some 215,000 tons in 1981. After years of calls, good dividends were now being paid to the patient shareholders. 1874 was the year of the crash in tin prices and the average price received by the mine fell to just over £56 per ton compared with nearly £80 in 1873. Nevertheless production of black tin increased from 301 tons in 1873 to 475 tons in 1874 so, despite the crash, the mine's revenue actually increased. Even when the price fell to below £37 per ton in 1876, a price that made mining sub-economic for most mines in the county, the mine continued to make good profits. It is difficult to relate prices from the 1870's with those of today. However an example was the price of South Wales coal that then cost £1 per ton delivered on the mine!

In 1875 a new lease was agreed with the Pendarves estate. The new lease also included extra ground to the west and south-west of the mine – this southern extension was on the western boundary of Wheal Grenville and it offered South Condurrow the opportunity to work the Great Flat Lode in depth in this area. Logically Grenville should have had this area as they could then naturally extend their workings westwards should payable ground be found. However the two adjacent mines were working minerals owned by different landlords.

In 1881 work was started on Marshall's Shaft. This shaft was designed to work the Flat Lode and the steep lodes that extended westwards into the new patch of ground acquired in 1875. By 'raising and sinking', the shaft was down to the 70 by the beginning of 1882. 'Raising and sinking' in this case meant that they had extended the 30, 42 and the 70 levels west from workings further east until they reached the projected position of the shaft. Sinking was then started from surface and at the same time a raise (a vertical or steeply inclined tunnel) was started from the 30 and driven up to meet the shaft sinking from surface. Once holed through the shaft would be equipped (timbered and the guides and ladders put in) down to the 30-fm level. The whole process would then be repeated – sinking below the 30 and raising up from the 42. This method is far quicker than simply sinking from surface as the shaft can be attacked from two or more points. A secondhand 26-inch winder from West Chiverton mine was erected at the end of 1881 and later, in 1886 a 60-inch pumping engine from Wheal Jane was put to work on the shaft.

Plate 3 shows Marshall's 60, date unknown possibly after the closure of South Condurrow. Note the old fashioned headframe behind the taller shears that were used to lower pump rods down the shaft. The rope from the whim is slack indicating that the cage has been landed on surface. The capstan rope from the shears has been pulled back and also is slack. The tramway from the shaft to the stamps ran away to the right.

Throughout much of the 1880's production ran at about 500 tons of black tin per year and the mine paid regular dividends. However, by the end of the decade, production was falling and in 1891 the mine failed to pay a dividend. Production was still coming from the old part of the mine but reserves here were almost exhausted and the bottom levels had been abandoned and allowed to flood thus throwing water down into Wheal Grenville.

The only hope for the future lay in the lower levels at Marshall's shaft. Unfortunately results here were very disappointing. By 1894, with tin prices again falling to sub-economic levels, the mine's production was only just over 200 tons. The tin price continued to fall and, with little mineable ground left, the mine was closed in 1896.

South Condurrow Mine – 1899-1903

The closure in 1896 was not the end of the story, for in 1899 the shareholders decided to reopen the western end of the mine around Marshall's Shaft.

An interesting glimpse into how a mine was run is given in the setting report of July 1900. They were sinking Marshall's Shaft below the 167 fathom level using 9 men who were paid £25 per fathom sunk. The 167 fathom level was being driven at £10 per fathom. They had 4 pares or gangs of tributers working on Engine Lode at 13/4 in the £. From this we can see that their sinking and driving was being done on 'tut work' at a fixed rate per fathom of advance. The stoping was on 'tribute' and the pare of men were paid on the value of the ore that they mined. 13/4 in the £ meant that they got 13/4 (66.7p) for each pounds worth of ore that they mined. Whilst the system prevented waste it was difficult to control and each pare's ore had to be weighed and sampled separately.

Tribute rates would be set at regular intervals and the rate became a binding contract between the mine and the miners.

The tribute system became increasingly difficult to justify as mines began to use more efficient and cheaper methods of drilling, tramming, hoisting and dressing. One result of going away from tribute was a fall in the grade of the ore hoisted, due to less selective mining, but the ability to handle larger tonnages did not necessarily mean a fall in tin production.

By September 1900 a new dry, or changing accomodation, had been built at Marshall's and the boiler house was being extended There were plans to modernise the old

Plate 3. Marshall's Shaft, pumping and winding engines

dressing floors by the installation of a Frue vanner plant. There were 32 men on tutwork, 18 tribute and 20 on surface. All this was costing money with no return. Investment continued with the erection of an air compressor at Marshall's and the building of a tramway from the shaft to the stamps.

Plate 4. 90-fm level South Condurrow – 1893

Finally, after two years work the first sales of tin were made in November 1901. However, the losses continued and the first sign of real problems came when it was announced that some shares had been relinquished. By June of 1902, 1076 shares had been relinquished[1] but over 5000 shares remained and the chairman believed that a further 400 would soon be relinquished. It was decided that they could not continue and decided to look for further capital. The company was reconstructed as a Limited Liability Company early in 1902.

The new company had very limited financial resources and, as a final trial, it was proposed to sink Marshall's a further 15 fathoms and to drive east and west to prove the shoot on the 193. It was also proposal also to crosscut south to Polgine Lode which was only worked to the 62. In July 1903 the directors reported:

> Unfortunately the anticipations found as to the payable volume of this section of ground have not been realised.

The company was wound up voluntarily and South Condurrow passed into history.

Early Student Activity

This is an appropriate point to look at some of the underground practical work done here by mining students before the mine ceased production.

J.C. Burrow took a number of photographs underground in South Condurrow in the early 1890's. These give us a glimpse of the working conditions of the period, and perhaps help us further to understand why there was a need for the School to have its own mine. **Plate 4** was taken in the 90-fm level in 1893. Here a group of students are seen 'hard at work' in a

1. In a Cost Book company the shareholders were liable for any losses and these were funded by 'calls' from these shareholders. When a shareholder reached the point where he was not prepared to fund further losses, in the absence of a buyer for his shares, he could relinquish then by handing them back to the company. This threw a greater burden on the remaining shareholders.

stope on the Flat Lode. In practice a large group like this would never work in such close proximity. It is important to imagine what this scene would have looked like illuminated solely by candle light. The men have removed their 'hard hats' because it would be easier to work without them, their candles being stuck on convenient rocks to illuminate the work. The 'hard hat' of the period was made of felt stiffened by resin and looked a bit like a bowler hat. There was no liner as in a modern safety helmet and most men wore a cotton skull cap under their hard hats. The hats offered a limited amount of protection – their main use being to carry a candle. In the centre foreground is an excellent illustration of hand drilling a blast hole. To the right a man is pushing an end-tipping wagon; note the double flanged wheels which were a loose fit on the axles and were better able to follow the poorly laid rails.

The naming of levels in Cornish mines can be confusing. Usually the datum was adit level with subsequent levels, normally referred to in fathoms, below adit. So here the 90 fathom level was 90 fathoms below adit. Much of the early underground prospecting was done by driving adits in search of lodes or by driving them along lodes which had been discovered from surface exposures or pitting. Once payable ore had been found shafts would have been sunk below adit following the lode in depth. Hence the

Plate 5. 90-fm level South Condurrow – 1893

Plate 6. Reinforcing an overhang, South Condurrow - about 1895

sunk below adit following the lode in depth. Hence the section of the shaft above adit would often be vertical with the rest of the shaft below adit being inclined following the dip of the lode downwards. This can lead to a further complication. Some mines measured the depth of their levels down dip, that is on the incline. In this case the 90-fm. level would be shallower than in a mine that measured the depth vertically.

The geometry of mine workings, being long and narrow, make them difficult places in which to take photographs. Photographers tend to take their underground photographs in the wider sections of the mine. This can give a misleading impression as to the normal size of the workings. **Plates 4** and **5**, taken on the 90-fm level in 1893, are examples of this, taken in a particularly wide section of the Great Flat Lode with a large posed group of students.

Here the student group are enjoying well earned pasties whilst being recorded for posterity. Notice the miner's dial on its tripod in the foreground. This instrument was the forerunner of the theodolite of today. It consisted of a magnetic compass to give magnetic bearings and a vertical alidade or circle for measuring the vertical angle. The old miner to the right, with the beard, is Capt. Nicholas Temby who was in charge of mining practical work. We shall meet him again later.

Plate 6 shows a group working to support an overhanging block of rock. Judging by the very rough state of the workings this was an area that was no longer in regular production. Note the water bottle and candles carried by the man nearest the camera.

Plate 7. Clearing a run, possibly 60-fm level. The date is uncertain but has been captioned previously as 1895. Mining conditions here are very dangerous as there is a weak roof or hangingwall. The large rocks on the right of the picture will have previously fallen from the roof.

Plate 7. Clearing a run, 60-fm level – 1893 or 1895

Chapter 2

Development – The Early Years

The mine was to be far more than just a place of instruction. It was intended to run the mine on a semi-commercial basis using a permanent staff of miners aided by crews of students, the tin produced going part way towards the cost of operating the mine. The day to day running of the mine was handled by the Mine Foreman, Captain Nicholas Temby. He had been with the School for sometime and appears in several of the underground photographs taken in the early 1890's. All the mining, surveying and mineral dressing elements of the School's courses would henceforth be taught at the mine.

The mine was administered by a Mining Committee which was a sub-committee of the General Committee (or Governors) of the School. The Committee included J.J. Beringer – the Principal and Arthur Thomas[1], soon to become the manager of the famous Dolcoath Mine. William Thomas, was the Mine Manager and responsible for all operations at the mine.

The School took over the area around Engine Shaft which had been sunk to 40 fathoms below adit or 400 feet from surface. There were shaft stations at Shallow adit, Deep adit, 10-fms, 20-fms, 30-fms and 40-fms which was also shaft bottom. This offered the opportunity to work both William's Lode on all of these levels below adit and the Great Flat Lode which had previously been worked to a certain extent on the 30 and the 40-fm levels. William's Lode meets the Great Flat Lode at 58 fms below adit. The lode had also been worked from William's Shaft, vertical to adit and on the underlie to the 20-fm level or 280 ft from surface. **Figure 4** is an idealised north-south cross-section showing Engine Shaft and the connections to the Great Flat Lode and to William's Lode. To the south is King's Shaft which had been sunk on the underlie of West Basset Lode. King's Shaft did not feature in the School's re-working of this part of the mine

On surface the existing steam winder (whim) and sundry buildings were included in the purchase price of £325. The engine house on Engine Shaft had, apparently, already been demolished. There were no ore dressing facilities. The underground mine had to be re-opened, and surface facilities built from what was, in effect, an abandoned mine.

Re-equipping the Mine

For a small college like the Camborne Mining School to take on a mine on this scale was a massive undertaking and is probably unprecedented in mining education.

1. The Thomas family were to be involved in the affairs of the Mining School for over 100 years. Their connection started with Josiah Thomas then the manager of Dolcoath. He was a driving force behind the establishment of the School and also was the first Chairman of the Governors. On his death in 1901 he was succeeded as manager of Dolcoath by his son Arthur Thomas who was also to be a Governor of the School and was, for a number of years, a 'fill-in' Principal in the 1920's. He was later succeeded as a Governor by his son Len Thomas, a mining engineer who had studied at the School, and who, in time, became the Chairman. On Len Thomas's death his son Major Treve Thomas joined the Governors and remained so until the University of Exeter merged with the School in 1993. Treve Thomas's son, Tristan, also studied at the School.

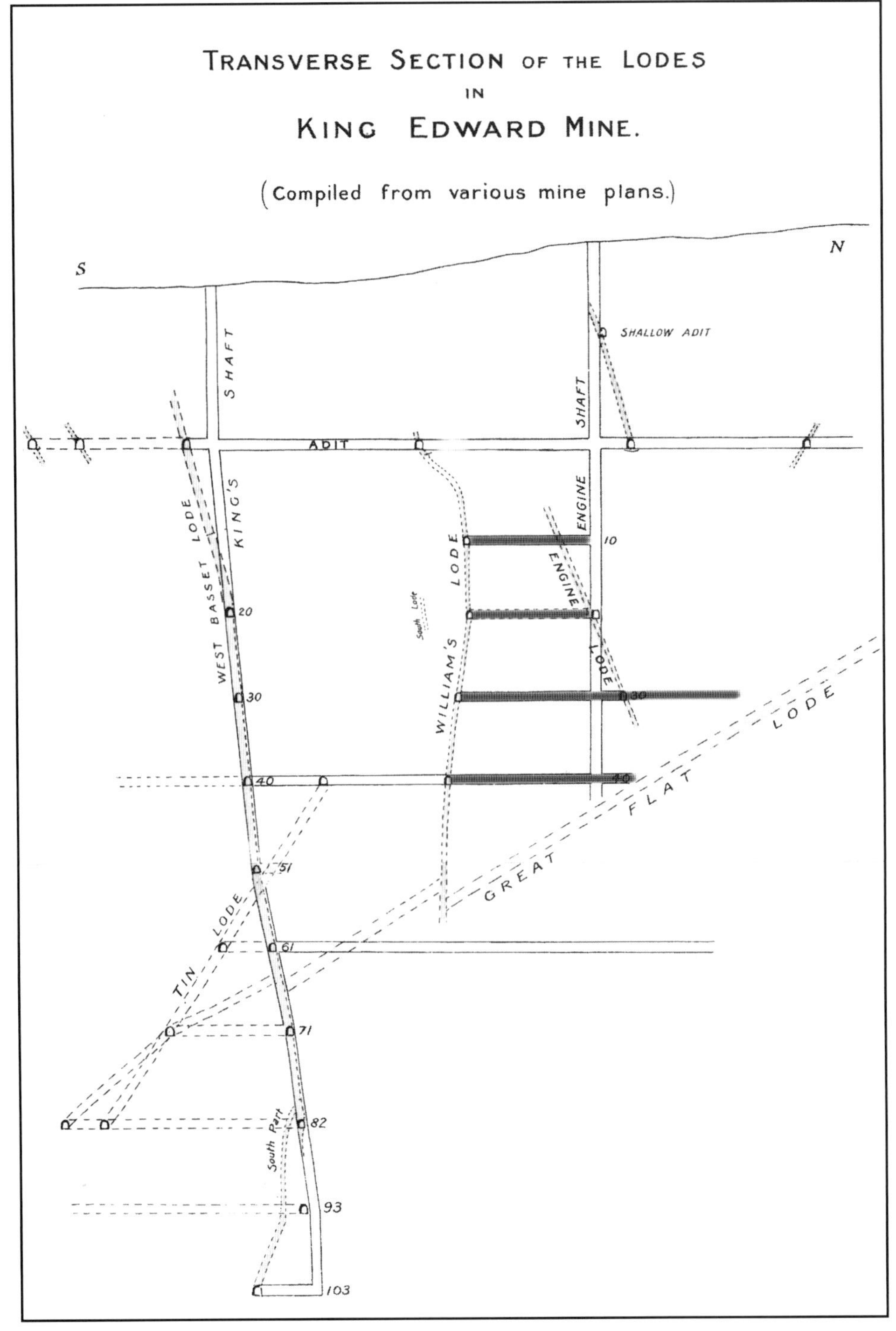

Figure 4. Geological Section

With the site came the old beam winding engine and boiler, a dilapidated building described as 'spalling shed' and the miners' 'dry' or change-house. The Mining Journal of March 1903 described the refurbished dry thus:

The changing-house is fitted up in a style which might, with advantage, be copied by other mines. Instead of drying with hot air the central tube is one long boiler, supplying hot water for 8 or 9 plunge baths which occupy two of its sides, whilst a second or smaller pipe supplies cold water. In addition to plunges there are 2 shower baths, so that in this respect – and it is a vital one – the students of the Camborne School enjoy very special advantages. **(Plate 8)** *Nor is the well-being of the practical miners who share the underground work of the students lost sight of, for their changing house too, is fitted with hot and cold water, of both of which there is abundant supply.*

The South Condurrow count house, blacksmith's shop, and assay office were retained by the old company but no doubt William Thomas managed to have some use of the blacksmith's shop. The single boiler alongside the steam whim, though an old type, was reported to have been in fairly good condition at least it was good enough to have been accepted by the insurance company to operate at 40 lbs. pressure.

The re-equipping of the mine was started immediately with the

Plate 8. Miners' dry – about 1899

Plate 9. Air Compressor – 1899

purchase, from Holmans Bros. of Camborne, of an 18 inch x 12 inch[2] compressor capable of driving two rock-drills. This was erected in a new building **(Plate 9)** close to Engine shaft. A little further down the site a new drawing or survey office had been completed by the middle of 1898. **Plate 10** is an early view of the Survey Office looking north-east. William Thomas can be seen standing by the corner of the building apparently 'supervising' a student surveying. Note the pieces of a Cornish pump lying in the foreground. These pieces had, presumably been pulled out of Engine Shaft during its re-opening.

Plate 11 is a general view inside the survey office taken about 1899. This appeared in the 1899 prospectus and the managers of the School must have liked it so much that it appeared on and off until 1932! Perhaps things did not change much. **Plate 12** was taken later about 1903 – note the modern theodolite on the table with an older instrument on the tripod behind the subject. The safe, in the background stands in the Survey office to this day (2001).

2. 18 inch stroke with a 12 inch diameter steam cylinder

By March 1898 The old head-gear on Engine Shaft, as well as that at King's Shaft, had been taken down. Engine Shaft was found to be in a very poor state. The timber at the collar had decayed so completely that the shaft at surface had collapsed so as to throw it 2 to 3 feet out of the vertical. The collar was completely re-timbered some 50 feet or more from surface, and the shaft made perfectly vertical down to the 400 ft level. The dimensions were 13 ft. 0 ins. x 5 ft. 6 ins. inside the timbers. In the western end of the shaft the old skip-roads were stripped out and a new double cage-road constructed. Adjoining the winding compartment was a new ladder road from top to bottom. There were suggestions of overflow pipes in connection with the water supply for a Pelton wheel which it was intended to fix at the 400 foot crosscut. It is not immediately apparent where the water was going to come from. Presumably any spare water in the dams could have been put to use to generate electricity. It never happened.

King's Shaft was stripped of all ladders, skip guides,

Plate 10. Survey office, headframe, compressor house and spalling shed – about 1899

Plate 11. Interior of the Survey office – 1899

Plate 12. Inside the Survey office – about 1902

headgear and other useful material which they thought would be of use. Subsequently the shaft was used as a dump for the stuff excavated when the foundations for the mill engine and stamps were taken out.

Shortly afterwards a new head-gear was erected at Engine Shaft from drawings made by Mr F. B. Lewis, who was described as an old and well known student, though one presumes that the 'old' referred to the fact that he was a past student.

Plates 13 and **14** illustrate clearly how a headframe was erected before the days of the mobile crane. Notice the chain, anchored to one of the head frame legs, running along at ground level towards the left. The lifting action of the sheerlegs will tend to drag the whole structure off to the right and this chain prevented that. Notice also, in the left foreground, the concrete foundation that would later take one of the backstays from the headframe. Beyond the headframe is the newly erected compressor house, whilst in the distance from left to right are: the old South Condurrow Stamps; the weighbridge and assay office with the tall chimney. Furthest to the right is the count house, blacksmiths shop with the double doors and, the miners' dry. **Plate 15** shows the work well advanced. The timberwork to support the guides has been completed, the ropes put on and the cages suspended in the shaft.

The head-gear, which was 44 feet in height to the centre of the pulleys, was constructed to carry

Plate 13. Erection of headframe over Engine shaft – about 1898

Plate 14. Erection of headframe over Engine shaft – about 1898

the detaching gear for Humble's patent hooks, which were to be fitted to the cages. A Humble hook or detaching hook, which is still used in mining today, is fitted between the end of hoisting rope and the suspension gear on the cage. In the event of an overwind this 'hook' engages in a catch plate near to the top of the headframe. As it does so the top of the 'hook' opens and the rope is detached from the top of the hook. The rope then continues on and over the sheave wheel whilst the cage is left suspended safely in the headframe. Prior to this invention, a full overwind often resulted in a broken hoist rope and the cage falling away back down the shaft.

By November 1899 the winding engine had been overhauled and the cages were ready to start. Unusually for an old shaft, Engine Shaft was vertical and thus could accommodate what can be described as modern cages. **Plate 16** shows the cage in the western compartment with 3 miners about to be lowered into the mine. Note the simple safety bar across the width of the cage. No doubt this was the norm and the instruction would be 'do not stick your arm out' – today's Mines Inspector would shut the place down immediately if he saw something like this. It would have been a damp and rattly ride down to the working levels, the cage only illuminated by the light of a candle or two, that is if they had not blown out or been put out by a drop of water. The old man on the left is the cage-man and he is about to ring the cage away and is holding the 'knocker line' or bell wire in his right hand. Pulling this line would transmit a signal to the winder driver via a mechanical linkage. A code of signals would have been set up by the mine manager with copies posted in the winder house, at the shaft top and at each level underground. When not used for man-riding, the cage was also used to hoist ore. The cage would take one loaded wagon.

In the east cage compartment the cage has been removed and replaced with a bailing skip. A bailing skip is used to hoist water from the shaft bottom or sump. They are usually used in an emergency to supplement the mine pumps on occasions when they cannot cope with the water entering the mine. Bailing water is inefficient, expensive and also ties up valuable hoisting capacity. In this case, it was installed in order to provide additional water for the boiler and the dressing plant – more of which later. The fact that we can see both the cage and the bailing skip on surface together tells us that the winder was only working the rope in the western compartment. It is normal practice, and far more efficient, with a twin drum winder, to run the two conveyances in balance, i.e. one cage going down whilst the other is being

Plate 15. Newly erected headframe – 1899

hoisted up. From an engineering point of view, in this case, the weights of the cage and skip would just about balance out.

A rock-drill fitting shop was provided, and four large granite blocks, one of them weighting 4 tons, had been obtained for surface drilling practice. A competent rock-drill foreman, who for many years superintended rock-drill contracts at West Frances, Wheal Basset, Tregurtha Downs, Polberro and others had been engaged

The general view, **(Plate 17)**, which shows the above works completed, appeared in the 1899/1900 prospectus and is contemporary with a student's plan produced in 1900, **Figure 5**. At about this time the mine became known as King Edward Mine, named after King Edward VII who came to the throne in 1901 on the death of Queen Victoria. Quite when the name was adopted is not known but the first reference to it appears in April 1901.

Amongst other additions to the buildings on the mine were a new material house, a vanning room and a new room for the class in timber-cutting. (Appendix I) With the large increase in the number of students in

Plate 16. Cage on surface at Engine Shaft. Note the bailing skip in the east compartment - about 1902

practical mining and surveying, it was reported in the October 1902 CSM Magazine that the accommodation in the 'dry' was altogether insufficient. The Survey Office also had become too small to accommodate the numbers taking that course. A proposal was made to extend the western end of the office to the road. This would have added about 20 ft. in length to the building. There was no money for this extension and it was finally carried out in the late 1960's. An approach was made to the Manager of South Condurrow in an endeavour to get permission to use a room in their offices. It was also noted in 1902 that there was a possibility of South Condurrow changing hands. It will be recalled that it had restarted in a small way in October 1899, and the School wished to rent the office used by the South Condurrow Company.

On surface, prospecting north up the hill, just north of the Carn-y-Mor farmstead (now demolished) had found the outcrop of the Flat Lode. Further prospecting using a series of costean pits two fields to the east again found the back of the Flat Lode. North Shaft was then started some 300 feet still further east

Plate 17. General view looking north – about 1899. L-R foreground: Survey office, Blacksmith's, Spalling shed, Compressor house, Headframe, Boiler house. L-R background: South Condurrow Count House, Blacksmith's shop, Miner's dry and Carpenter's shop

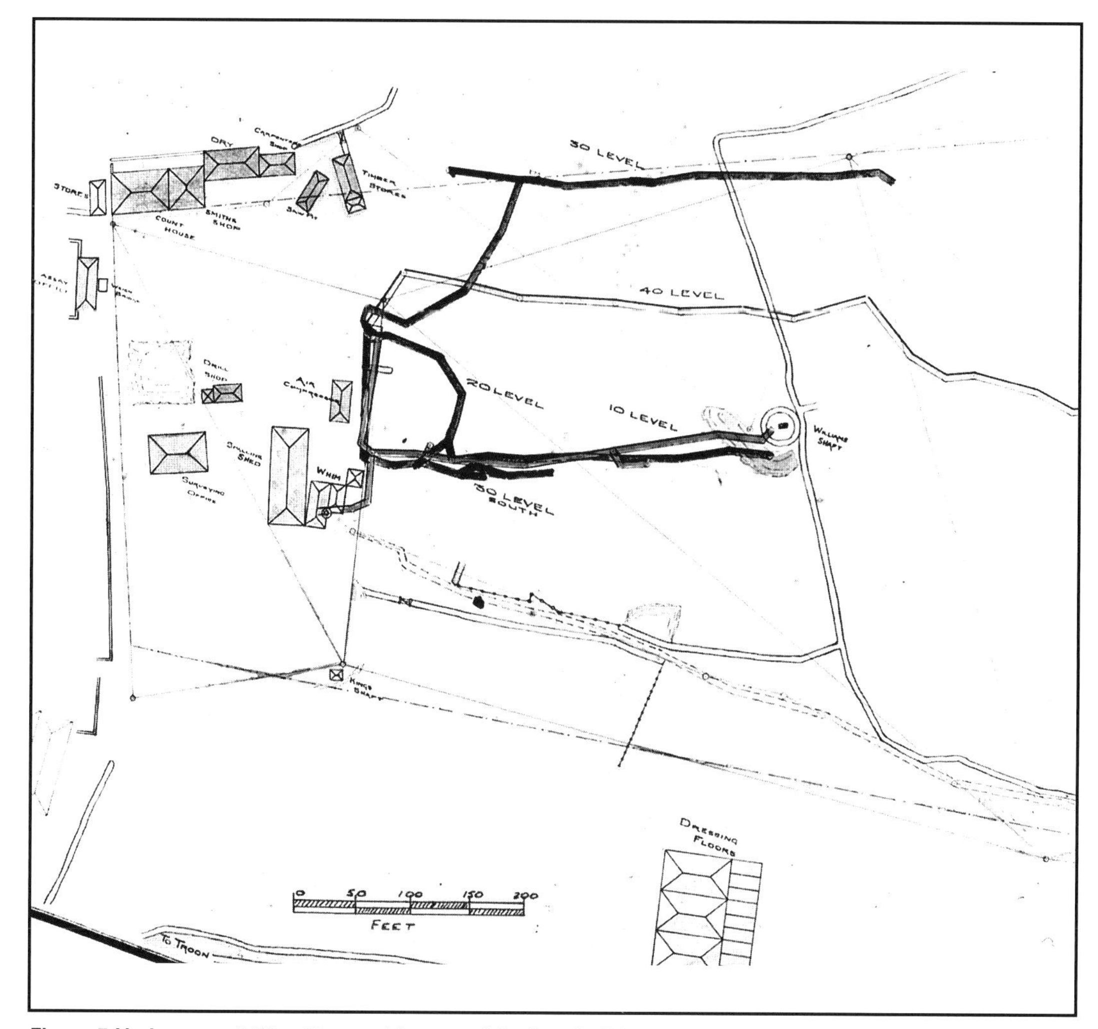

Figure 5 Underground Mine Plan and Layout of Surface Buildings – 1899

and, sunk vertically, intersected the Lode some 50 feet from surface. Work then continued sinking down dip.

At the beginning of 1902 William Thomas visited the Royalton mine, which lies to the north of Goss Moor, to buy some wagons for use underground. Back in Camborne, proceedings were taken against three boys who stole materials from underground, including some dynamite. The magistrate sentenced them to 6 strokes of the cane each. If that was not nuisance enough, in April the General Committee, as always watchful over expenditure, complained: *as to the price of miners' candles at 4/6 per dozen compared with 3/6 on other mines.* It turned out that they had been specially ordered by William Thomas, as he did not think that *the soft candles used by some mines would, if purchased for the School, be so economic as had the candles that they were now using.*

In March 1903 the Mining Journal described the mine as:

...about as pretty a working model of what a Cornish Mine should be as could be desired. The buildings, all of which are substantial and well lighted and heated, consist of a surveying office, with all the necessary space and appliances for 45 students, assay office fitted with furnaces, muffles etc. for 10 students, smith's, carpenter's, fitting and timber cutting shops, stores, general offices etc.

Also in 1903 Wheal Grenville purchased the western part of South Condurrow. Mr. Thomas, always keen to improve King Edward, reported:

Now that the South Condurrow Mine has been purchased by Wheal Grenville, it is necessary that some arrangement be made with the latter, as soon as possible, regarding the Count House and detached buildings, etc. No doubt an arrangement is possible whereby, at a nominal cost, several improvements may be at once effected. For example it is desirable to put one of our men to live on the mine, to arrange for a students' mess-room and a cycle house, and make some other little improvements on the mine.

It was later noted that the buildings had been purchased for £100.

Chapter Three
Underground at the Mine

In this chapter we look at the re-opened underground workings, as seen through the eyes of the photographer.

Having drawn a supply of candles, a group of miners and students pose for a photograph **Plate 18** before going underground at Engine Shaft around 1903. Note Captain Temby in the background, wearing a bowler hat, with some of his permanent staff of miners (with the candles) and perhaps mature students. In the foreground the young students look anything but happy. The cheerful old man on the lower left is one of the surface workers.

Engine Shaft was vertical to 40 fms below adit (27 fms) giving an overall depth of 67 fms or almost exactly 400 ft. Confusingly, reports at that time state depths in both fathoms and feet. In Cornwall it was very unusual to record depths in feet. It could be that William Thomas, aware that the rest of the world used feet, tried to introduce this more modern way of reporting depth. However, for this narrative we will use the traditional fathom.

The shaft provided access both to the Great Flat Lode and to the steeply dipping William's Lode. Figure 4, which we saw in Chapter Two, is a north-south cross-section of the workings around Engine and King's shafts. Deep adit was the first level of importance. This was found to be in poor condition and, as far as we can establish, the crosscut was only re-opened as far as William's Lode. Below adit William's Lode was accessed by crosscuts south on

Plate 18. Group of students and miners at Engine Shaft – circa 1903. Capt Temby in the background in the bowler hat

the 10, 20 30 and 40-fm levels. The Flat Lode was found to have been accessed by crosscuts on the 30 and 40-fm levels, driven north from the shaft to the lode. From the ends of the crosscuts the lode had been opened up by drives driven eastwards following the lode, the longest drive being on the 40.

On the Flat Lode the 40-fm level was cleared, and an inclined skipway put in from the 40 to 60-fm level. The reason for installing this incline is not known as no further reference has been found of work below the 40-fm level. **Plate 19** a ladder road between the 40 and the 50-fm levels is probably the skipway referred to. The bearded miner seen here and in Plates 7 and 20 could be Josiah Luke who was one of School's underground supervisors. The Flat Lode was one of Cornwall's major orebodies, but the payable ore tended to occur in shoots separated by sub-economic zones. It was a wide structure that resulted in large excavations where it was extracted. We can see here that the lode must be all of 9-10 ft. thick and the relatively flat dip of the lode is very apparent.

Plate 20. This huge excavation, captioned "The Cathedral", was on the 50-fm level below adit. The unusual width of the Flat Lode here could have been the result of faulting caused by one of the steep lodes, such as William's Lode, intersecting the Flat Lode. Just such an intersection is shown in Figure 4

All mines need a second means of access as an escape way in the event of an emergency and to provide through ventilation. To achieve this, William's Shaft to the east was reopened. The shaft had been filled within 15 or 20 feet from surface. In November 1902 it was reported that:

Our first trouble in clearing the shaft was with a dead horse or two! We have cleaned the shaft 45 fms below surface and are now sinking it. We have put in a new ladder road from surface and enclosed the top of the shaft in a house.

The Mining Journal reported:

William's Shaft has been re-opened and the 20 level on Williams lode completely cleared and re-opened and a new tram road connecting the two shafts. The 20 and 30 are at present choked, but are being re-opened and the spiling[1] and timbering operations on these levels making excellent practice for the students.

Plate 19. Ladder road between the 40-fm and the 50-fm levels – about 1901

1 Spiling is a method of advancing a tunnel through very soft ground or a collapse. A number of timbers, or 'spiles' are driven into the collapse along the line of the tunnel around its edge. Some of the broken rock is then cleaned out and timbers are put up to support the spiles which act as a roof and walls of the tunnel. Further spiles are driven in and the process repeated as the tunnel advances.

Plate 20. The Cathedral 50-fm level below adit – about 1901

In a steeply dipping lode like William's, the initial exploration would have been by driving horizontal tunnels, or drives, following the lode. Where it was found to be worth mining, the lode would normally have been removed by overhand stoping. The method was as follows. The miners would first have blasted down the ore in the roof of the level. After 2 or 3 blasts upwards – having stripped off the roof to a height of perhaps 7 ft. above the original roof – a timber bulkhead would have been constructed as an artificial roof to the level so that mining could continue on upwards, and the level used again as a through haulage way. Drilling and blasting would continue on upwards with the miners standing on temporary platforms built on cross timbers, or stulls, jammed between the walls of the stope, the broken ore falling to the bottom of the stope. The blasted ore would have been hand-sorted and any waste or low grade material left on the floor of the stope above the bulkhead. Over time, as the stope became higher, a considerable amount of waste could be left in the stope which would remain after the area was abandoned. In time the timber bulkhead would rot and eventually some or all of the broken waste would run down choking the level. The subsequent reclamation of old levels, even today, is a slow, difficult and dangerous job.

On William's lode the Journal states:

The old workers did a lot of work, mainly on tribute, on this lode from the 20 upwards, but on the discovery of the Flat Lode 30 or 35 years ago the smaller lodes were neglected and this lode has not been worked for some 25 years before we purchased the sett.

The old workings eastwards on William's Lode were opened up and driving commenced on the 10, 20, 30, and 40-fm levels east. Tram roads were laid as the levels were cleared. The new rails were T-section and

Plate 21. Students surveying on the inclined skip road – about 1903

Plate 22. Stoping on the Flat Lode just above the 40-fm level – about 1904

were fitted together with fish plates and bolts.

Once the mine workings had been re-opened they were surveyed and the results plotted on the mine plan. **Plate 21**, taken in 1903, shows a group of students surveying in an inclined skip road somewhere on the Flat Lode.

As the shaft and haulage ways were opened up it was possible to start producing ore from the Flat Lode. Most of the work done here was on, and immediately above, the 40-fm level. The tin values in the lode here were very patchy and the 'old men' had left most of it in-situ as being sub-economic to extract.

Students would have been exposed both to the traditional method of hand-drilling and the more modern rock-drilling. Hand-drilling was still very common in Cornish mines at this time. Despite William Thomas's efforts to be up-to-date, it was used extensively at King Edward, as the mine only had 3 rockdrills. **Plate 22** shows a group of students stoping on the Flat Lode just above the 40-fm level.

One of the three machines was photographed about the same time also on the 40-fm level. **Plate 23**. The rockdrill seen here is mounted in a cradle on a jack-bar and arm as it is too heavy to hold by hand. The machine which is rotary percussive (similar to today's domestic hammer drill) is fed forward by a manual screw feed which the miner is operating with his left hand. The lad behind the machine is holding a water spray which was supposed to help to control the dust. A high speed machine like this will crush

Plate 23. **Rock drill with dust laying spray. Location unknown – probably Flat Lode**

the chippings in the hole to a fine dust before it has a chance to escape out of the hole. This very fine dust, only a few microns in size, if it contains silica, can cause silicosis – a potentially fatal lung condition. Today, the modern rockdrill has the facility to pass water through the drill steel to the end of the hole thus wetting any dust before it has a chance to escape into the atmosphere. The tank on the right is a pressure tank. This will have been filled with water and a compressed air line attached to it, the air pressure forcing the water out of the tank to the spray nozzle[2].

After the holes had been drilled they were charged with gelignite and safety fuse. Safety fuse was invented by William Bickford in Tuckingmill in 1831. Originally it was designed to set off charges of black powder[3] – the fuse, once lit, safely transmitting a flame down to the charge in the hole. Safety fuse consists essentially of a core of black powder wrapped in layers of waterproofing and cotton binding. It burns at a constant speed (normally 30 sec/ft). Gelignite is not flame sensitive and requires a detonator to set it off. A detonator is a small metal tube containing a tiny amount of a flame sensitive high explosive. The detonator is crimped onto one end of the fuse and this is pushed into one of the sticks of gelignite when the hole is being charged. **Figure 6.**

Figure 6. Nobel Explosives Company Invoice – 1909

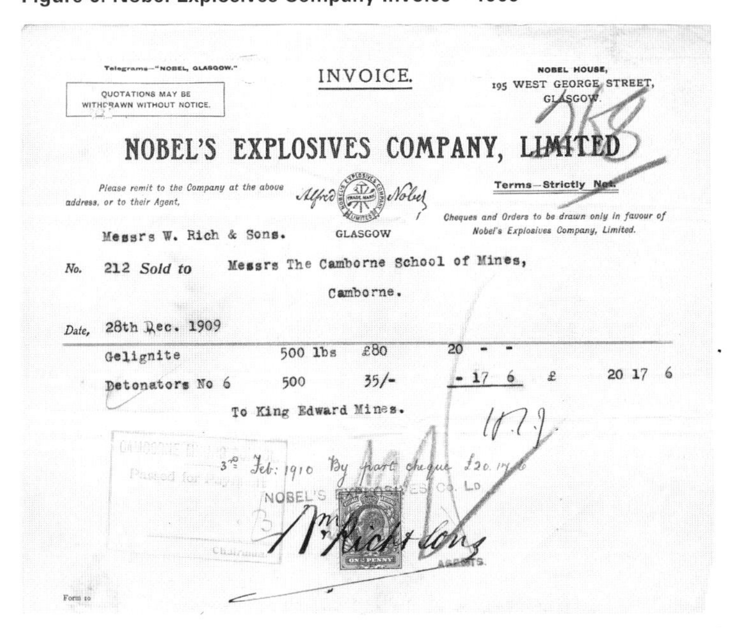

Telegrams—"NOBEL, GLASGOW."

QUOTATIONS MAY BE WITHDRAWN WITHOUT NOTICE.

INVOICE.

NOBEL HOUSE,
195 WEST GEORGE STREET,
GLASGOW.

NOBEL'S EXPLOSIVES COMPANY, LIMITED

Please remit to the Company at the above address, or to their Agent,

Alfred Nobel

Terms—Strictly Net.

Cheques and Orders to be drawn only in favour of Nobel's Explosives Company, Limited.

Messrs W. Rich & Sons. GLASGOW

No. 212 *Sold to* Messrs The Camborne School of Mines,
Camborne.

Date, 28th Dec. 1909

Gelignite	500 lbs	£80	20 - -		
Detonators No 6	500	35/-	- 17 6	£	20 17 6

To King Edward Mines.

3rd Feb: 1910 By part cheque £20.17.6

NOBEL'S EXPLOSIVES Co. Ld

W Rich & Sons

AGENTS.

Form 10

2 This type of pressure tank would be illegal today for, if it had a weakness, it could burst. These are long gone from the mining industry although some were still in use at the Great Rock Mine in Devon in 1963.

3 Black powder is the mining term for gunpowder.

When it is time to blast, the fuse is lit and providing the fuse is long enough the miner has plenty of time to retire to safety before the explosive detonates. **Plate 24** shows a miner lighting safety fuse on the 40-fm. level, about 1903. This excellent close-up shows the typical clothing and resin impregnated felt hard hat worn by miners at that time. Note the ball of clay wrapped round the candle.

This sticky clay, which came from a pit on the side of St Agnes Beacon, was used to attach the candle to the hat when the miner needed both hands free.

As soon as the fuses were lit the miners would retreat along a safe exit. **Plate 25** shows a team of miners retreating down a

Plate 24. Lighting safety fuse, 40-fm level – about 1903

Plate 25. Retreating after lighting the fuses. 40-fm level Flat Lode east of Engine Shaft – about 1904

stope above the 40-fm. level, having lit their fuses.

After the blast the dust and fumes would be allowed to clear before the men returned. They would first make the place safe checking and barring down any loose rock. Then the ore would have to be helped down the slope of the lode to the lower level and then shovelled into wagons. **Plate 26** illustrates this well as three men pause for the photographer. This was probably taken on the 40-fm level Flat Lode about 1904. The wagon, once filled, was pushed along the level to the shaft. At the shaft the wagon was pushed into the cage, secured, and then hoisted to surface. On surface an elevated tramway had been constructed linking the headframe with the mill. From the cage the wagon could be trammed directly into the mill to be dumped into an ore bin above the stamps. This was a highly efficient layout that eliminated any unnecessary handling between the stope and the mill. Returning to Plate 26, of particular interest is the student with his hands on the wagon. He appears to be an African. Whilst CSM was beginning to develop an international reputation, the presence of a student from Africa in Cornwall at this time must have been almost a unique occurrence. So far we have not been able to identify him.

Plate 26. Filling a wagon. Probably on the 40-fm level Flat lode – about 1904. Nationality of the student in the centre is unknown

Little or no level development was done on the Flat Lode. The ore values were not good and once the mine was established most of the work was concentrated on William's Lode. The old men had done little work on this lode so the levels were driven east to open up new ground for extraction. By the end of 1902, the CSM Magazine reported:

> *Underground several additional points of operation have been started. On William's Lode five ends are being driven east and one west. Two winzes are being sunk on this lode and William's Shaft will be sunk below the 20 level at once. The sinking of the new North Shaft has been resumed on the back of the Flat Lode and efficient ventilation, by means of a small furnace*[4] *has now been secured here.*

4 'Furnace' - no description of this survives. It is thought to have consisted of a metal pipe put down the length of shaft with a small furnace fitted to the bottom. Once lit the fire would create a strong airflow up the pipe (or chimney) to surface which in turn would draw fresh air down to the shaft bottom.

Here on the 30-fm. level, William's Lode, two men are hand drilling the face of the level **Plate 27**. The man on the right steadies the drill steel for his mate, who is striking it with the hammer, and then turns it between blows. A portion of the lode has been blasted out of the roof above the miners' heads, possibly for sampling purposes. The ladder behind the men probably leads up into the stope being worked above the level. Note the men's hats dropped below the ladder and the single candle stuck on the wall.

A similar activity is taking place on the 40-fm level east, **Plate 28**. Whilst all of the levels below adit at King Edward were in granite, ground conditions were patchy and some areas required support. Timber support like this is the exception rather than the rule in Cornish mines in the granite.

The picture that William Thomas would have liked to present is that of a Climax rockdrill **Plate 29** in the end of the 30-fm level William's lode. This machine has a water spray attached to the machine, a refinement on the hand held spray seen in Plate 23.

Along with horizontal development, vertical connections had to be put in between levels. These were needed for ventilation and also to sample the lode between levels. Raising up is easier than sinking or winzing down. In a raise, whilst access is more difficult, all water and broken rock falls out

Plate 27. Driving the 30-fm level east

Plate 28. Driving the 40-fm east by hand

Plate 29. Rockdrill in use with water spray. Date & location unknown – probably on either the 30 or the 40, William's lode

Plate 30. **Sinking a winze below the 30-fm level William's Lode – about 1904**

Plate 31. Sinking a winze below the 30-fm level

of the raise and even drilling is easier as the blast holes are self-cleaning. Burrow and Hughes both had an eye for a well composed photograph and were keen to exploit back lighting in their images. **Plate 30** taken by Hughes, shows 3 men working a windlass when sinking a winze below the 30-fm level, about 1904. This may look a crude system but exactly the same method was still in use at the Geevor tin mine in 1962.[5] The absence of a wagon or any broken rock by the tracks, the

5 In 1962, Tony Brooks worked as a student at Geevor cleaning just such a winze, which was being sunk below 9 level.

Plate 32. Tramming – probably on William's Lode – about 1906

empty bucket and the drill steels leaning against the wall indicate that the men are about to lower materials prior to drilling. Further along the drive as the tracks swing away to the right is an ore chute from a stope. A similar operation is shown in **Plate 31**, again sinking a winze below the 30-fm. level. Note how soft and friable the broken rock is in the pile near to camera.

King Edward was a shallow mine and temperatures underground would have been cool. The shirtless young miner pushing an empty wagon back from the shaft **Plate 32**, had clearly been doing some hard work.

Because of the steep dip of William's Lode, the broken ore from the stopes could be loaded through chutes directly into wagons. **Plate 33**, an obviously posed photograph, demonstrates apparently how easy it is to fill a wagon. The miner has removed one of the flow boards and propped it against the end of the chute and is encouraging the broken ore to run into the wagon. In reality these little chutes, often referred to as Cousin Jack chutes, are difficult to control. The rock tends either, to jam somewhere up in the chute,

Plate 33 Pulling a chute – circa 1904

Plate 34. Party underground – date and location unknown. Probably about 1904

or comes pouring out uncontrollably all over the wagon and onto the ground. Chutes of this type continued to be used in the Cornish mines up until the late 1980's. Note the compressed air pipe suspended from the roof.

A quick photograph for the family before heading back up to the shaft to catch the cage. The group, in **Plate 34**, place and location unidentified, could be visitors as they look remarkably happy and clean. On the 40-fm level shaft station a group of students and their instructor wait their turn for the cage **Plate 35**. Note the cage in the background about to be rung away to surface.

Back on surface, **Plate 36**, a visiting party of Royal Artillery Officers from Woolwich are seated on the steps in front of the Survey office about 1903. This is probably one of the last photographs taken of Capt Nicholas Temby, sitting in the doorway, as he died on the 12th of March 1904.

Plate 35. Waiting for the cage on the 40-fm level – about 1902

Plate 36. Visiting party of Royal Artillery Officers from Woolwich. Taken in front of the Survey office – 1902/3

Chapter Four
Building the Milling and Dressing Plant

As mine development progressed, a committee was appointed to consider the laying out of the dressing floors. The science, perhaps better described as the art, of tin dressing had altered little throughout much of the 19th Century. New methods and equipment had been developed abroad towards the end of the century, particularly in the United States, but Cornwall was slow to adopt them. This was due to a number of reasons including a natural conservatism from an industry that had led the world, the attitude 'why change something that works' and the Cost Book system where profit was distributed immediately leaving little in reserve for capital works. Years of low tin prices had also put off shareholders and management from investing in an uncertain future.

This is an appropriate point to look at the problems that faced the tin dresser and the equipment and methods available in 1900. Tin is not found naturally as metallic tin. The mineral that was mined for tin in Cornwall is cassiterite or tin oxide. This has to be smelted in order to produce tin metal. Tin ore can be defined as a mixture of cassiterite and other minerals that can be mined profitably. Where found in commercial quantities, the percentage of tin metal in the lode is low, typically running between 1 and 3%. The bulk of the minerals in the ore, such as quartz, chlorite etc. are of no commercial value and must be removed, as far as is practicable, before the final product, or concentrate, is sent for smelting. Frequently associated with the tin oxide are other metallic minerals containing copper, iron and arsenic. Whilst these may be potentially valuable they also must be removed from the concentrate before smelting. A final tin concentrate will contain about 60% tin metal.

One of cassiterite's characteristics, important from a dressing point of view, is its high specific gravity of 6.9. This can be compared with most non-metallic minerals that occur in the ore which have a specific gravity typically in the range 2.5 – 3.0. It is this high specific gravity differential that is used to separate tin from its associated waste minerals, in so-called gravity separation. The metallic minerals, mentioned above, are sulphides such as chalcopyrite, a common ore of copper which has a specific gravity of 4.2, iron pyrite 4.9 and arsenopyrite 6.1, tend to concentrate with the tin.

The size of the individual particles or crystals of cassiterite in the ore is important, as this will dictate the extent to which the ore must be ground in order to liberate the cassiterite from the associated waste minerals. This will be an imprecise process as the tin oxide mineralisation will be in a range of sizes. Thus to release all of the tin the ore will have to be ground very fine indeed, and the result is that most of the coarser tin will be over-ground. The finer the tin oxide the more difficult it is to concentrate by gravity methods. The stamping or grinding of the ore and all further concentration processes are done in water, and the very fine or 'slime' tin oxide, rather than settling, tends to be washed away with the waste. This is the problem – how fine should it be ground? The solution is to crush the ore only sufficiently to liberate the coarse cassiterite. This should then be removed before the remainder of the ore is further crushed to liberate the finer tin oxide.

The South Condurrow dressing floors that lay just to the west of the present access road into the site was typical of the traditional Cornish dressing plant that would have been found on the tin mines of the late 19th Century. It is shown in **Plate 37**, the photograph being taken about 1895. The crushing was carried out by a battery of Cornish stamps. The Cornish stamp battery consisted of a long cylindrical cast iron stamps barrel that was rotated by the steam stamps engine. In this case there were stamps on either side of the stamps engine, but we can only see the eastern side in this illustration. Set in the barrel were a number of lifters, usually 5 per head, that picked up the stamps as the stamps barrel revolved. The stamp itself, weighing about 750 lbs., consisted of a rectangular cast iron metal head attached to a metal stem. On the stem was a collar by which the stamp was picked up by the lifters. The ore, which had been hand cobbed into pieces about 2 ins. in size, was fed by waggons into chutes that led down to the coffer box where the crushing was done. At this point water was added. In front of the coffer would be either a punched metal screen or a wire screen, the perforations controlling the maximum size of the product from the stamps.

Plate 37. South Condurrow stamps and dressing floors – about 1895

The pulp from the stamps was run into strips, long narrow pits which acted as holding tanks and also allowed the pulp to thicken before passing to buddles. The strips in the photograph were probably behind the raised masonry walling in front of the stamps. Below the wall, to the left, can be seen at least 2 rows of buddles. The de-watered pulp that had settled in the strips would be re-pulped using water to get the right consistency before being led to the buddles in wooden launders. Such a launder can be seen running down the plant behind the first buddles, with a feed into each buddle. It is not known what was in the small wooden buildings.

The concentrates from the above processes would still contain heavy minerals such as the sulphides of copper, and iron. These were removed by roasting or calcination. When heated, or 'burnt', to the right temperature the sulphur burns off as sulphur dioxide gas leaving copper and iron oxides. These are light and friable and are easily washed away from the cassiterite which is unaffected by the calcination. Arsenic, in the form of arsenopyrite behaves differently. The resulting oxide sublimes as a vapour at high temperature and, where there was a significant amount of arsenic in the concentrate, this was collected as white arsenic by cooling in long flues. Where there was a small amount of arsenic, it was allowed to escape into the atmosphere to the detriment of the local countryside. The calciner at South Condurrow was off the photo to the left. The stack by the engine house served both the engine's boilers and the calciner. Little or no arsenic occurs in the Flat Lode area and no attempt was made to collect that which existed.

The last operation before the commercial concentrate is ready for the smelter is known, in Cornwall, as 'tossing and packing'. In this, the concentrate, previously cleaned from impurities as far as possible, was fed slowly into a tub part filled with water. The concentrate was kept in quick rotation by an eddying movement set up by several men using long-handled shovels, the stirring being stopped after the whole of the charge had been delivered into the tub. The tub was then struck with a mallet, whilst the particles were settling. The high grade material was found in the bottom of the tub and lower grade at the top The lower grade material was skimmed off and retreated, the rest was bagged, weighed and sold.

In the background of Plate 37, peeping over the skyline to the left of the stack is the roof of the mortuary chapel which still exists.

The Dolcoath mine in Camborne was one of the few that had modernised and invested in much new plant. William Thomas was fortunate to have his cousin, Arthur Thomas, on the managing committee. It was intended to instal a fully operational tin dressing plant employing labour saving methods and the most up-to-date equipment. The plant that was erected at King Edward reflected much of what was the best from Dolcoath with a layout at Engine Shaft very similar to that which existed at Valley Shaft at Dolcoath.

The flowsheet finally decided on is shown in **Figure 7**.

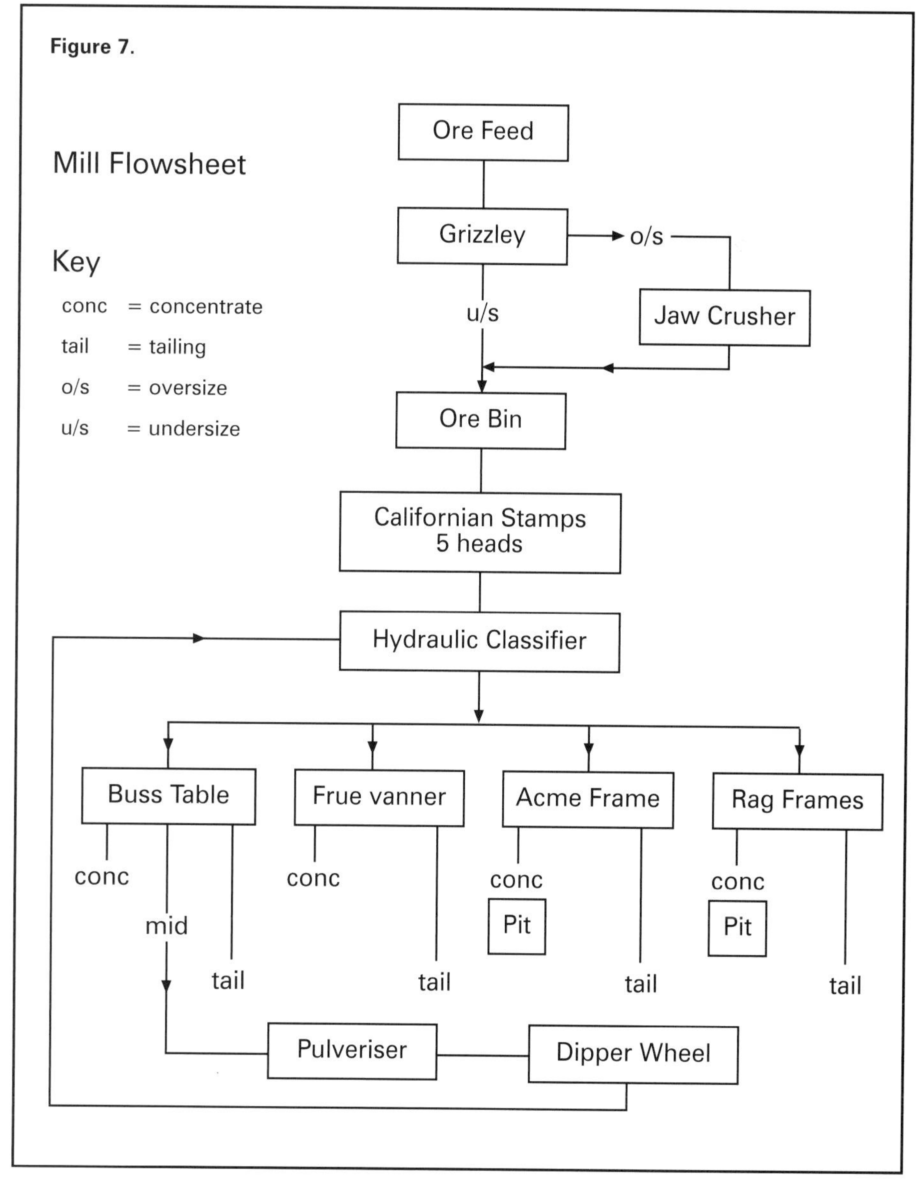

It was planned to drive all of the mill machinery, including the stamps, with a horizontal steam engine. **Plate 38** shows work on the excavation for the engine foundations. Behind is the compressor house. A 90 H.P. compound steam engine, was ordered from Holmans of Camborne. **Plate 39** shows the completed mill engine – note the high quality of the finish to the interior of the building. The pulley to the right of the flywheel carried the belt that passed down into the mill

Plate 39. 90-hp Holman mill engine – belt drive down into the mill on the right – about 1902

Plate 38. Newly erected headframe, compressor house to the right. Excavation for the foundations of the mill engine in the foreground – 1899/1900

and drove the mill machinery. The door at the far end of the building led into a small lobby and thence into the compressor house.

In September 1900 the West Briton newspaper recorded:

The County Council a few weeks ago, made a further grant to the Camborne School to enable them to proceed with the erection of stamping machinery, and it is understood that this will be put in hand immediately.

The funding was not just for stamping but also for the associated dressing plant.

A five-head set of Californian stamps, complete with small jaw-crusher, was purchased from Fraser & Chalmers, this set having been exhibited at the Paris Exhibition of 1900. Californian stamps were the latest in stamping machinery, The first to be installed in Cornwall were at Dolcoath in about 1893. The purchase of the set for King Edward in 1900 coincided with installations at the Royalton and Bunny Mines in mid-Cornwall, and at Gunnislake

Clitters mine in the Tamar valley. The stamps were a nominal 850 lbs. weight with a drop of 8 ins. operating at about 100 drops/minute.

Plate 40, taken in 1974, shows the detail of the stamps mortar box. The screen has been removed and is leaning against the mortar box foundation. The five heads have been blocked up using props that are out of the picture at the top. Each stamp head has a replaceable shoe. The shoe has a tapered stem that is simply wedged into the bottom of the stamps head using pieces of scrap wood as packing. The stamping action drives the shoe in to the end of the head. The shoe can be removed by driving a steel wedge into a slot on the side of the head above the point that the shoe taper terminates in the head. Ore was fed into the mortar box from behind the heads along with a steady supply of water. The water feed pipe, here seen disconnected, runs across the top of the mortar box above the maker's name. The steps on the left lead up to the feeder and cams platforms.

Plate 40. Californian stamps – 1974

Plate 41. Details of the Californian stamps cams – 1974

Plate 41, also from 1974, was taken on the cam platform and shows the details of the cams. The collars on the stamps stems are wedged in position by steel wedges. The collars would be periodically adjusted to compensate for the wear of the heads. The heads of a stamp battery like this would wear about 1 lb. of metal for every ton of ore stamped. The lifting action of the cams caused the stems of the stamps to rotate slightly with each lift guaranteeing even wear of the stamps heads.

One of the advantages of modern stamps such as Californians, as opposed to the traditional Cornish stamps, was that they were normally used in conjunction with an automatic feeder driven by the up and down motion of one of the stamp heads. The further the head moved up and down the more ore was fed into the stamps mortar box. Conversely, if the ore was being fed in faster than it was being stamped, the head's drop would decrease and the feeder would feed at a slower rate. This feed rate could be adjusted to give a chosen level of ore in the mortar box.

Plate 42 shows a survey team of students plus a number of 'helpful' onlookers attempting to survey in the base of the stamps. The building on the right is the compressor house and the larger building in the centre is the mill engine building. The very large timbers on the ground are probably part of the stamps framing awaiting erection. The vertical posts are part of the arrangement used to erect the stamps framing – note the rudimentary scaffolding on the mill engine room

Plate 43 was probably taken on the same day as Plate 42. This is an unusual view of the mine as it is the only one that we have found taken looking west up the valley. Note how the beam winding engine house has been built on an elevated foundation. This is uncommon and it is suggested that this design was adopted in order to keep the hoisting ropes clear of the ground. During the main period of working of the mine the headframe on Engine Shaft would probably have been much smaller than that put up by the School. The south end of the 'spalling' shed can be seen projecting out beyond the boiler house.

Plate 44 was taken a few weeks later. The stamps have been erected and the building is being put up around them. The old 'spalling' shed has been demolished

Plate 45 shows the framing of the stamps building almost complete prior to sheeting. Compared with Plate 42 it

Plate 42. Surveying the base of the Californian stamps – probably 1901

Plate 43. King Edward seen from the east – about 1901

will be noticed that a porch has been added to the compressor house and a door access made into the back of the mill engine room. This is the same door as can be seen in the photograph of the mill engine. One puzzling item in this photograph is what appears to be a massive concrete foundation just to the left of the compressor house but clearly separate from it. Perhaps there had been plans to purchase a different stamping plant – plans that were changed when the Californian battery became available? Judging by the decrepit state of the whim house with boarded up windows, facia board missing and pieces of the weather boarding on the south end of the boiler house falling off, repair of this building was not a priority. The photograph is captioned as being taken by H.A.B. He cannot be positively identified, but is probably Henry Arthur Burrow (son of J.C. Burrow) who studied at the School between 1901 and 1905.

Plate 44. Erecting the stamps – about 1901

The stamps produced a product that contained: a small volume of free cassiterite, a larger volume of material that contained a mix of tin and waste and the largest proportion, perhaps 80%, that was just waste. All of these three products would be in a range of sizes

It had been recognised for some time that gravity separation worked better if, after stamping, the material was sized. The different concentrating machines were designed to work on a particular size of range of material.

It was planned to classify the crushed tin ore into three sizes: rough sands; fine sands and slimes using an upward current classifier.

The rough sands were to be treated on a Buss Table. The Buss table was very new, the patent being applied for in 1899. The first Buss tables to be installed in Cornwall were at Gunnislake Clitters, the second at King Edward and third at the Nanturras mine near Marazion. This table along with the Wilfley table were widely used in Cornwall in the early years of the 20th century. They were later replaced by the more successful James table, the first of which was installed in Cornwall around 1912.

Plate 46 shows the Buss table seen here to the right of a Frue vanner (described later). At the top of the photograph are the stamps with the steps leading up to the stamps feeder and cam platforms.

The Buss table consisted of a rectangular table or deck, about 12 ft long and 4 ft wide, inclined across the shorter axis, and supported on a number of flexible legs. The deck was fitted with a number of thin wooden strips or riffles, of varying length set a few inches apart, tacked onto the deck along its long axis. The longest strips were near the lower edge of the table and the shorter nearer the feed or upper edge.

Plate 45. Stamps in course of erection. L-R Stamps, Mill engine room, Compressor

Plate 46. Interior of mill – about 1903

At the feed end a vibrator 'shook' the table along its long axis with a gentle push along the table followed by a sudden jerk backwards.

The ore bearing pulp from the classifier was fed onto the deck at the top at the vibrator end of the table. The pulp flowed down across the deck the solid particles becoming trapped behind the first riffle. The shaking action would tend to make the dense tin particles sink to the bottom of the layer with the lighter waste particles coming to the top. The particles containing both waste and tin, the middlings, would stratify between the tin and the waste.

As more material became trapped behind the riffle the lighter material would overflow down the table to the next riffle and the same process would be repeated. The jerking action of the vibrator would drive the dense material still trapped behind the riffle along the table away from the vibrator. The best tin would be caught by the upper riffles with a succeeding fall in grade as the overflow material worked its way down across the table. As the concentrated material moved along the table behind the riffles it was washed with trickles of clean water. This washing helped further to clean the concentrate and to wash the waste down across the table.

Finally the small volume of high grade concentrate was collected in a box at the end of the table. The middling was discharged below the concentrate and, at King Edward, was taken to the pulveriser for further grinding. The waste, which was by far the largest product from the table was discharged into the tailings launder and eventually went into the river.

One of the great advantages of the shaking table over buddles and vanners was that a considerable proportion of the waste was discarded at this stage of the concentration process.

The middle products from the Buss table went to a tube mill or barrel pulveriser as it was known in Cornwall. The barrel pulveriser was patented in 1880 by Messrs. Michell and Tregoning, and the first one was made by the Bartles Foundry, Carn Brea. The unit that was installed at King Edward was also made by this foundry. A pulveriser is, in essence, a revolving cylinder part filled with steel balls or scrap. The fine sands and water are fed in at one end and the tumbling action grinds the material. At King Edward, **Plate 47** the discharge of the milled product was elevated by a dipper wheel, which can be seen beyond the pulveriser, to the main stamps dipper wheel and thence back to the classifier for re-sizing.

Plate 47. Barrel pulveriser used to re-grind the middlings from the Buss table. Probably about 1905

The bulk of this finer material would go onto the Frue vanner. King Edward was the first plant in Cornwall to utilise a pulveriser in a closed circuit.

The fine sands from the classifier's second outlet were passed to a Frue vanner. Plate 46 The Frue vanner consisted of an endless rubber belt 4 or 6 ft. wide and 12 ft. long supported on rollers. The belt mechanism and its drive was supported on a number of short legs. The upper surface of the belt, which is slightly inclined, travels upwards towards the end at which the pulp is fed on. An eccentric at the side of the machine communicates a lateral oscillating movement to the belt. The pulp is fed onto the belt from a distributer situated near to the top or discharge end of the belt. The heavy mineral grains sink on to the belt, and, owing to their weight, are able to pass under a series of jets of water, and to be carried over the end of the machine with the belt where they are washed off into a trough of water situated below the belt. The lighter particles of waste are carried back down the belt by the wash water in a direction opposed to that in which the belt is travelling.

Frue vanners were first introduced at Dolcoath in 1898. They were an immediate success and over 70 were installed at Dolcoath to be followed by major installations at Wheal Grenville and Basset Mines Ltd. They were most effective on fine material and worked well where mines were stamping down to a very fine size. The disadvantage, compared with a shaking table, was that they did not produce a middling and tin in the tailing discharge had to be treated by traditional means to recover any residual tin. Some survived at South Crofty and at Geevor up until the late 1950's where they were used, as at King Edward, to treat the fine material.

The slimes were to be treated on an

Acme table that was an improved form of revolving slime frame. The conventional Cornish round frame consisted of a slowly rotating circular concave wooden table. The feed was introduced from a fixed curved launder running around the outside of the table for about two thirds of its circumference. The slime then gently flowed down across the table towards the centre. The dense tin tended to stop on the table and the lighter waste and the water flowed off in the centre. As the table rotated it moved out of the feed area and into an area where clean water was fed onto the table from a similar fixed launder to that which fed the slime. This water further cleaned the concentrate on the table before it was swept off at the end of the rotation by a jet of water and a brush. The Acme table was a development of the conventional round frame where the deck was split – the concentrate from the first part being elevated by a small dipper wheel before being retreated on the second part of the frame. This modification was something of an over complication and they were not a lasting success. The conventional frame continued to be used on stream works until the 1960's. The Acme table at King Edward is shown here in **Plate 48**. The old tin dresser is giving the outer part of the frame a clean. In the background, beyond the Acme table, is the small dipper wheel elevating the product from the outer section of the frame to the feeder on the inner section. The concentrate from here was passed to the settling pit in the foreground. Note the Frue vanner and Buss table in the background as in Plate 46.

Plate 48. Acme table with on the left the Frue vanner and Buss table, – about 1904

The dressing floors were designed to keep within a tight budget. Financial stringency was the watch word throughout the life of the mine and the School's Governors kept a very close watch on expenditure.

During the summer of 1902 the engine-house and the house built to accommodate the crushing and dressing plant were completed, at least as far as the exterior went. Inside the engine-house the engine itself was finished and steam had been tried, with satisfactory results on the engine.

Ore grades underground varied but, by 1902, W. Thomas considered that they were in a position to return sufficient ore yielding 30 to 40 lbs tin to the ton to keep the stamps working 16 hours a day. Whilst this may sound impressive, a 5 head battery of this type (850-lbs per head at 100 drops per minute) would only be capable of stamping 10 – 12 tons per 16 hour day.

One vital unit was missing – a calciner. It was planned to build this a little further down the site below the dressing floors. It was reported in October 1903 that:

Ground being cleared to erect a reverberatory furnace with drying hearths and to recovery tanks for acid treatment.

The ore contained a certain amount of copper. Once calcined the copper would become soluble in dilute acid. The copper in the resulting liquor could be recovered by putting it in a tank with scrap iron. The copper being precipitated out as 'cement' copper.

With the final closure of South Condurrow in 1903, the School was able to take over the count house, smith's shop and carpenter's shop. These were the stone buildings running across the top of the site. Also acquired at the same time were the weighbridge, the brass-casting house and, of vital importance to King Edward's dressing floors, the ponds and watercourses. The stamps started during the summer of 1903. **Plates 49** and **50**, also from 1903, were probably taken on the same day. Plate 49 was taken from the tramway that took the ore wagons to the South Condurrow Stamps. Of note in this photograph is the miner pushing a loaded wagon along the elevated tramway connecting the headframe on Engine Shaft with the stamps. The cage can be seen in the headframe behind the miner. The cloud of exhaust steam above the mill engine room, part obscuring the back-stays of the headframe, indicates that the mill engine is running. In the distance, to the right over the Survey Office, are the whim and pumping engines on Fortescue Shaft, Wheal Grenville.

The London and West Country Chamber of Mines summed up the progress of the surface works:

Despite the derogatory remarks made by a Redruth tradesman, that the installation at King Edward Mine included 'machinery that was rather to be avoided than recommended,' there is very little, if anything, to find fault with in this respect. Criticism of this description generally emanates from those who, knowing nothing of mining, know less of technical education, and were it not that they are calculated to do harm, should be beneath notice. It is,

Plate 49. L-R workshop, headframe and tramway to the mill – note the miner pushing a tram, mill engine room, stamps building. The 3rd section of the mill and the calciner are yet to be built

therefore, necessary to point out that in the installation at the King Edward (or Camborne Students') Mine it has been necessary to aim at one distinctive dressing plant of a particular type as to secure a combination of different types illustrative of those most in use at the present time, and yet forming one complete and workable plant. In this direction, Mr. W. Thomas and his staff have achieved a highly creditable success. Dealing only with the dressing of tin ores they have a Blake stone-breaker; Fraser and Chalmers' Californian stamps; Challenge ore-feeder and Frue vanner; three Luhrig classifiers; one Bartle's improved pulveriser, latest type, working on roller bearings, and belt driven from centre of barrel; one Acme table (Holman's patent); and one Buss table and several automatic samplers. This reads like the mixed contents of several machinery catalogues, but in spite of that fact the entire dressing plant forms one complete whole, in which the central idea has been to avoid handling, to classify before concentration, to learn value of produce, and to avoid loss in tailings. This is technical education as far as the exigencies of the situation will permit. That it is the perfection of dressing is not contended, but that it is calculated to help the student towards the determination of that important problem is beyond question.

Plate 50. Survey office, mill and winding engine looking east – early 1903

This was not enough for Mr. Thomas who tried to persuade the Mining Committee to purchase a pilot scale magnetic separator and dynamo (ex the Polmear Mine, St. Austell) it was rejected on the grounds of cost.

The first concentrate (3 tons) from King Edward Mine was sold at the end of 1904 for £73.00 per ton – a good price considering that the concentrate contained about 2% native copper and that the average price at that time was £76.00 per ton.

Chapter Five
Consolidation

The winter of 1903/4 was particularly wet and support problems were experienced in both shafts necessitating extra timbering in the upper sections. The Wheal Grenville pumps were struggling to cope with the additional water and, in February 1904, Wheal Grenville requested the use of the water skip in Engine Shaft. The water skip was designed to hoist water from the 40-fm level for dressing purposes, and any water removed from the shallow levels would lessen the load on the Grenville pumps. In view of the water problems experienced by Wheal Grenville, the management later approached the School with a proposal for installing an electric pumping plant at King Edward. Mr. Thomas's estimates for the proposed pumping plant were £1,225, of which he expected that Wheal Grenville would pay some £700. There was more to this than just pumping as King Edward had no electrical equipment at all, and the School had been criticised on this count. It was also hoped to use some of the excess power for magnetic separation. Mr. Thomas was asked to inform Wheal Grenville that the Committee were prepared to go ahead with the scheme. Unfortunately the plan was not implemented.

The weather also had an adverse effect on the winder house stack which now had a tendency to shed bricks in windy weather. Under some wind conditions it was difficult to keep the boiler fire going. Mr Thomas obtained a quote for rebuilding the top 12

Plate 51. Surface view – about 1904

feet and adding an additional 5 feet to increase the draught. A quote was obtained from a Mr. F Mitchell for completing the necessary repairs to the stack – his price, including materials was £15.

By the end of 1904 much of the surface works had been completed. The building with the square stack to the far right of the photograph, **Plate 51**, is the new calciner. Note the tramway running down from Engine Shaft to the flat area in front of the workshops. This tramway was later extended and used to transport rock from the small dump in the foreground up to the shaft. Here the wagon was loaded into the cage, hoisted up to the elevated tramway, and trammed into the stamps. Presumably this dump carried enough tin to warrant milling.

Plate 52 was taken at around the same time. The timber building on the left is the survey office with the open door that led into the bottom of the mill just beyond. Behind is the old steam whim and its lean-to boiler house. The new calciner, with its distinctive square stack, is on the right. Note the general mess and disorder of what was still, in effect, a building site. The stack on the whim engine house has been extended and the top of the stack is now sporting a corbel.

One drawback of the plant as completed was the lack of a calciner cleaner circuit. The roasted concentrate from the calciner would contain, along with the cassiterite, the oxides of iron and copper plus a small amount of waste material that would have been adhering to the particles of ore before calcination. The plant as constructed so far, meant ore milling had to stop whenever the burnt concentrate was to be retreated. Initially this was done using the existing Frue vanner and the Buss table. This arrangement was clearly unsatisfactory and Mr. Thomas recommended extending the dressing floors and installing an additional Frue vanner and Buss table which the manufacturers were prepared to supply at half price. A sub-committee, formed to examine the extension of dressing facilities, reported as follows:

The dressing floors are at present undoubtedly incomplete as there is no provision for dealing with the calcined concentrates which have therefore to be treated on the tables used for dealing with the pulp coming from the stamps, and consequently while the concentrates are being treated the stamp mill has to remain idle. This is inconvenient and expensive in itself, and moreover is not calculated to impress students with what should be a good practice in a dressing plant.

It was felt, that the expense of erecting further tables to

Plate 52. Newly completed calciner and stack with whim engine and boiler house behind – about 1904

deal with the relatively small tonnage of concentrates was not justified. However, the addition of a buddle was approved (**Plate 53**) instead.

The sub-committee went on to recommend that a small gold milling plant consisting of a tube mill and cyaniding plant be obtained as it was recognised that a majority of the students on leaving Camborne would obtain positions in gold mines. This advice was acted on and a small second-hand cyanide plant was purchased early in 1905. Back in 1899 there was an advertisement in the Mining Journal offering for sale a small gold plant consisting of three 8 ft. diameter steel vats 4ft deep, extractor box etc.. This description sounds exactly like the gold plant put in at King Edward. There cannot have been many plants like this available in the UK and one wonders whether they were one and the same. The mill was extended southwards towards the calciner, by the addition of a third bay, and the new buddle and ancillary work referred to above were completed. **Figure 8** shows the site at its completion about 1906 – note the tramway leading down from Great Condurrow to the old South Condurrow stamps.

This new extension also provided the opportunity to instal four sets of tossing gear, two of which also had mechanical packing gear. **Plate 54** This was excessive as one set of tossing and packing gear would have been sufficient for the small volume of tin concentrate that the plant was capable of producing.

By mid-1905 it had become apparent that the main steam boiler, which had been taken over from South Condurrow in 1897, needed to be replaced. It was being steamed at a low pressure, well under the 100 lbs. pressure for which the mill engine was designed. The problem with a new boiler, as with everything, was cost. A new Lancashire boiler would cost about £500 if it were to be sufficiently large to provide steam for the mill engine, the compressor and a new winding engine. The other alternative was a high pressure Cornish boiler which might be unable to run all the plant simultaneously but would only cost about £300. A Cornish boiler was ordered from the Cornwall Boiler Co. of Camborne.

Plate 53. Convex buddle in the mill – date unknown

1906 saw no slowing of progress at the mine and, in March, William Thomas reported that the new boiler was ready for delivery. He expressed concern as to the condition of

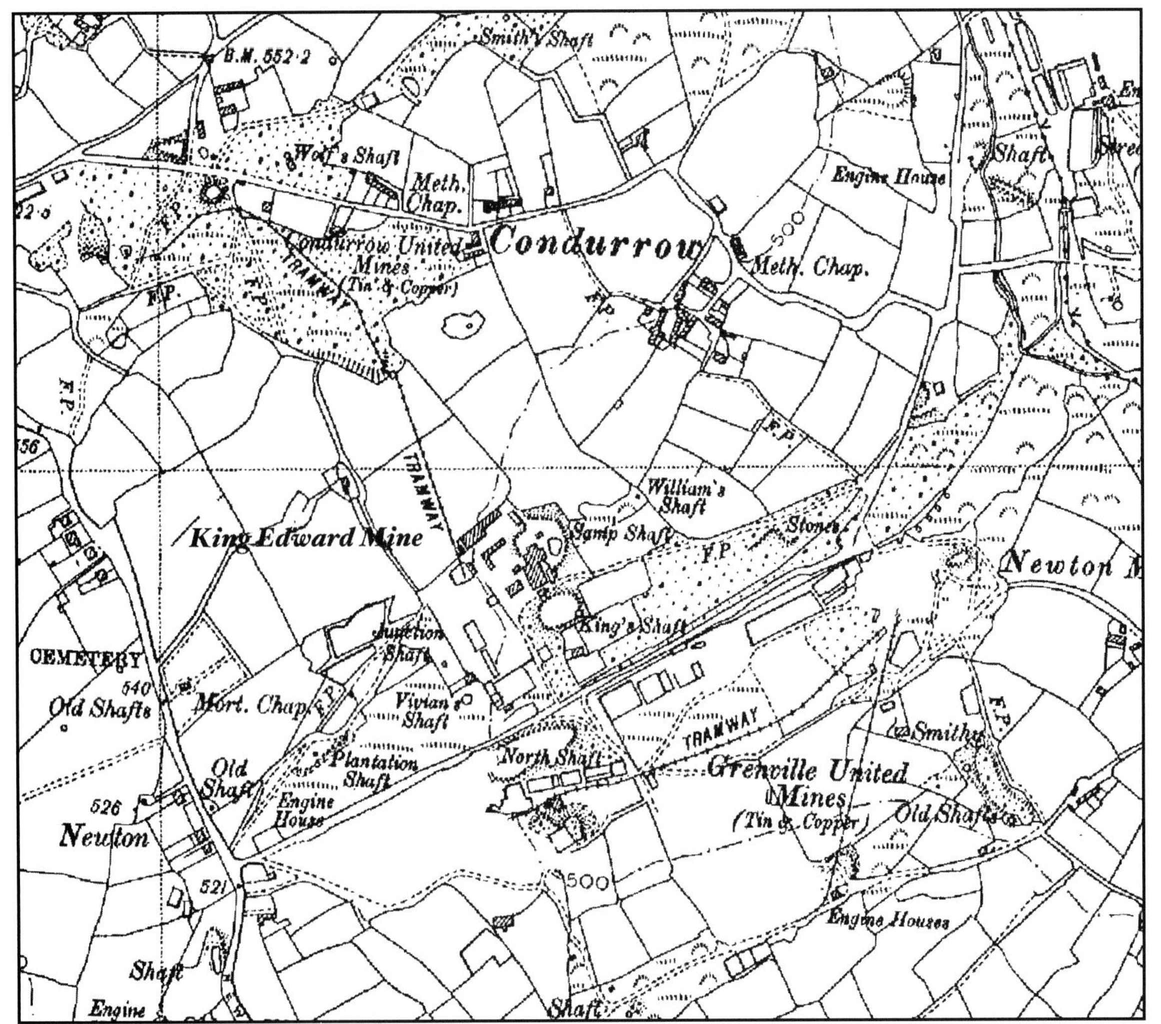

Figure 8. 1906 O.S. Map of Mine Area

the old beam winder for man-hoisting and suggested that it was the ideal time to replace the winder with a modern winding plant. Yet another sub-committee had been set up to examine the proposal to replace the old beam winder – nothing had been done and W. Thomas was evidently irritated by the lack of action, he reported:

It is unfortunate that the meetings of the mining sub-committee arranged to take place several weeks ago, have fallen through. The new boiler has been delivered on the mine but no steps can be taken to put it in until the Committee has decided what to do in the matter of the winding engine. I have already pointed out that the existing engine is old and defective. I wish now to add that I question the wisdom of continuing to hoist men with the engine as it is during the approaching session, and unless a new engine is provided or the existing engine put into good condition, I should ask for a special instruction from the Committee on the subject in order to remove some, at least, of the responsibility from my own shoulders. I find from Messrs. Holman Bros., that the price for such a pair of engines as we require would be about £300. I have reason to believe that if the matter is made known to some other good makers of winding plant, competition amongst them may considerably reduce this figure. I reckon that about £120 can be realised by the sale of the old engine and materials in the engine house. I strongly recommend that a new pair of winding engines be arranged for without any further delay, and that they be erected at the same time with the new boiler.

Finally his recommendations were acted on and, in September 1906, the Committee agreed to purchase a new winding engine.

Water, or lack of water, was a problem which William Thomas had not resolved. During the summer of 1906 they frequently had to stop the dressing plant because of the shortage of water. Hoisting water in the water skip had been tried but was hardly worth doing. Some additional water had been supplied from Marshall's Engine at Wheal Grenville but that had since been diverted to Wheal Grenville for dressing.

However, Willie Thomas's managership of King Edward was soon to cease. It had been reported that he had been doing outside work without permission and, on October 8th 1906, he resigned as Head of Mining and Surveying to become Manager of the Botallack Mine at St. Just, which was being re-opened by Cornish Consolidated Tin Mines Ltd. The loss of W. Thomas was a serious blow to the School as he had been largely responsible for the acquisition and development of the mine. At times his ideas and ambitions for King Edward had out-run the limited finance available and this had resulted in differences of opinion with the School Committee. Despite those periodic differences the Committee had, in the main, supported W. Thomas. Perhaps it was the right time for him to go and it would have been difficult for him to refuse such an offer. Having had the experience of successfully developing the little King Edward Mine from nothing, here was a chance to prove himself again on a far larger stage. Unfortunately when he left the stream of quality photographs of the mine, taken by his photographer friends, also ceased.

Plate 54. Kieves and Tossing and Packing Gear – about 1906

Before we meet Thomas's replacement let us have a look at a few of the characters amongst his staff and students.

One feature of the early School Magazines were the letters from past students updating those at home on their progress abroad or passing on advice. One from Australia advised students:

For outfit I shall recommend the following: Pyjama suits, flannel shirts with collars attached, 2 blue serge suits, 1 light and 1 thick, a light tweed suit, some pairs of grey flannel trousers, light merino socks and some Scotch hand-knitted socks to wear with heavy boots. Knickerbockers, riding breeches, norfolk jackets are not worn in the Colonies.......For preference P&O second class as they carry no third passengers and make a speciality of second class accommodation.....snakes are very plentiful, but will always get out of the way. In proceeding to kill one never attack it from the rear – always take them sideways or in front. A very slight blow breaks their back and they are powerless

Another, writing in 1897, recounted his experiences in British Guiana:

From what I hear, there are two seasons, the wet season and the rainy season. It is supposed to be the former when the river is low, as at present, but we had over 4 inches of rain in one night last week, and two inches in half an hour on Friday last, so if this is the wet season, what must the rainy one be like! I am at present located in a logie, a place with a paper roof, and wattled sides, it is the store for salt pork, beef, fish, etc and has been lent to me until my house is finished. When it rains the water drips on my hammock, and the whole place gets beastly wet.

At the height of the Boer War this obituary appeared in the Mining Journal of the 10th February 1900:

Sergeant Edgar Litkie, who had been a Camborne mining student. Litkie was leading a platoon of Thorneycroft's men when he was killed by an exploding shell, and subsequently buried on Spion Kop[1].

Edgar Litkie was born into a wealthy family who dealt in diamonds. His father, Valerian Arnold Litkie, was in business in Kimberley in 1873 as goldsmith, jeweller and watchmaker. It would seem that shortly afterwards he went to London, as in 1874 the journal Diamond News carried an announcement that he and his brother Emile had gone into partnership as diamond and general merchants and jewellers. The business in South Africa was handled by Emile, while Valerian looked after the London end.

The social standing of the Litkie family was clearly indicated by the reports of the wedding of Edgar's brother, Maximilian McKenzie Litkie, at Windsor in July 1895. Upwards of 200 guests were conveyed from London by special train from Paddington and the reception was held at Forest Park, Windsor. Edgar Litkie attended the wedding and of particular interest is the information that he had recently returned from Freiberg in Saxony, the celebrated town for practical mining engineering, in addition to schools of mining. From this it seems likely that he had some knowledge of mining before he came to Camborne.

Plate 55. Edgar Litkie

CSM records contain little information about him except for the Supplement to the Students' Register in which he is mentioned in the list of those who attended lectures in some or all subjects between 1887 and 1903, which indicates that he was not enrolled on a regular course. We know he was at the School in December 1897 as he put his name forward to fill a vacancy on the Students' Committee. The fact that on December 8th he was prepared to serve on the Students' Committee indicates that he was expecting to be at the School for some time to come. However the lure of gold was too much and in April 1898 we find him in Canada.

1898 saw the gold rush to the Klondike. Edgar Litkie and a local miner, Joe Thomas, decided that they would head for the Klondike as partners. Joe Thomas who had experience of gold mining in Africa and elsewhere, lived with his wife and family in Enys Road, Camborne. Litkie was a popular member of the student body. He gave a farewell dinner to his friends at the Commercial Hotel in Camborne, and the next day he and Joe Thomas (to quote a subsequent report in the Cornish Post & Mining News of May 19th 1898)

....were merrily escorted to the station by the Students' band, and left with hearty and kindly wishes... **Plate 55**

Ahead of them lay a journey across the Atlantic by steamer, followed by a train journey across America to

1 Anon. "A Student in the Klondike and South Africa" Camborne School of Mines Association Journal, 1997-98, p11-13.

one of the West Coast ports from where they would take ship for Alaska. They would have come ashore at Dyea, a few miles up the coast from Skagway, the gateway to the Klondike and the Yukon. Somewhere along the way the pair acquired a partner, a Mr. Chadwick, either on the passage to New York or when crossing America, and he was with them when they set out from Skagway. Skagway was a frontier town where stampeders, as they were known, had to assemble their supplies for the journey ahead. Entry to Canada was permitted only if the stampeders had rations and other supplies to make them self-sufficient for a year. From Skagway they joined thousands of other men making their way over the hazardous Chilkoot Pass into the Yukon.

The next report of them is on April 9th at Log Cabin camp on the Canadian side of the Pass. Log Cabin camp was where the stampeders rested after surviving the dangers of the Chilkoot Pass and checked their equipment and supplies before moving on to the next stage of the journey towards Dawson City and the Klondike. What happened there is best described in the words of the report in the Cornish Post & Mining News of May 19th 1898:

Klondyke kills a Cornish Miner. Mr Joe Thomas of Camborne dies at Log Cabin Camp. Sad end of Mr Litkie's Partner

Klondyke has claimed its first Cornish victim. The remains of Joseph Thomas, of Camborne, shrouded in the tent wherein he slept, lie in the frozen ground of Log Cabin camp, British Columbia. His grave was dug by his comrade and friend Mr. Litkie, a Camborne mining student, who toiled for eight hours before his melancholy task was completed.

On the 9th of April, Mr. Litkie, the deceased and another were camping in their tent at Log Cabin and all had colds, so thinking it would do his companions good, Mr. Litkie left them there and went out alone. On the Monday evening he had misgivings about something being wrong and on walking back to Log Cabin, he found Joe ill in the tent and unable to reply to what was said to him. About three miles from the camp was a Dr. Hatton and a nurse, Mrs. Egerton, who intends devoting herself to nursing work in the Klondyke. In response to Mr. Litkie's call the doctor and nurse came and the nurse and himself sat up with his chum four nights and days, never leaving him alone and with no time for meals. Mr. Litkie sent 60 miles for cow's milk, eggs and other nourishing food, but there was little or no chance for the unfortunate man. They were living in a tent, with snow and ice everywhere outside, and a cutting wind causing frequent variations in the temperature, so that the tent was sometimes like an oven and sometimes like an icehouse. The poor fellow's temperature was nearly all the time 105 degrees and sometimes more and at last double pneumonia culminated in death.

Worse was to come for Edgar Litkie. His family in London received word that the other partner, Chadwick, had also been taken ill and had died in hospital in Skagway[2]. No details are available on Chadwick, except confirmation that he did not book his passage through the same emigration agent in Cornwall as Litkie and Thomas. It has not been possible to trace their movements in the Klondike area or find out if they had any success in the goldfields.

After two years in the Yukon, Edgar Litkie went to South Africa. It may be that he had enough of the Klondike, but he had close family connections with South Africa. His uncle, Emile Maximilian Litkie was Captain and Paymaster of the Kimberley Rifles. With the outbreak of the Boer War this military connection may have influenced his decision to go there with the intention of enlisting. On his arrival in South Africa Litkie was desperate to join the colours, as is shown in the following extract from a letter to his father, written from Pietermaritzburg and quoted in the obituary –

Here I am, trying all I know to join some corps, but so far of no avail. It is very trying to idle here when one knows there is fighting going on so near to here. I still hope to succeed. I am determined to go to the front. Anyhow you will know I am doing my best to do my share to help Queen and Country.

He succeeded in enlisting in Thorneycroft's Mounted Infantry, an irregular unit raised in Pietermaritzburg on October 16th 1899 by Major A.W. Thorneycroft of the Royal Scots Fusiliers. Four months later, on January 24th 1900, he was killed at the battle of Spion Kop in Natal during General Buller's abortive attempt to relieve Ladysmith. He was buried in a communal grave along with 29 of his comrades and another 60 were wounded in the battle. Edgar Litkie died in South Africa only two years from the time when he and his friends enjoyed their farewell dinner at the Commercial Hotel, Camborne. In those two years he experienced more than many people do in a lifetime.

To commemorate the centenary of the Gold Rush the State of Alaska awarded certificates to descendants or relatives of those who were involved and whose connection can be authenticated. Mrs. Marion Rogers, Camborne-born grand-daughter of Joe Thomas, has been a recipient and has generously donated the certificate to the School of Mines where it is exhibited.

In the same year, W R Bateson wrote from Concordia Mine in Argentina, which was at over 12,000 ft. and a three day mule ride from the town of Salta:

This is the most lonely place possible, 120 miles from a town. Still I can be very happy here with plenty of work to do, otherwise it would be a bit lonely even for me, because the country around is practically a wind swept desert.

Another student from the same period was Harold Sharpley (1880-1928). He had an interesting pedigree in that his great-grandfather was Francis Basset (1757-1835), later Lord Dunstanville of Tehidy, whose monument stands on Carn Brea barely a mile from King Edward. It would appear that Sharpley's grandfather was the result of a relationship between 19 year old Francis Basset and Emily Martin, of a respectable Methodist farming family. Unfortunately Emily died in childbirth so we will never know whether they would have been permitted to marry.

Harold Sharpley was the son a of a Lincolnshire farmer, though his mother came from Camborne. No doubt it was as a result of his Cornish connections that he came to the School of Mines. He entered the School in 1898 and left in 1901 having obtained a First Class Diploma and the Silver Medal of the Cornish Miners' Association. He will have been at King Edward Mine with William Thomas during the early years of the mine would have witnessed many of the developments recorded by the photographs in this book.

Sharpley's subsequent career is typical of many mining engineers of the 20th Century. They travelled and managed mines the world over. Sharpley's first job was with the Anglo-German Mining Syndicate on their properties near Macri in Asia Minor. From 1904 to 1909 he worked for Messrs. Bewick Moreing & Co in charge of the surveying department on some of their mines in Western Australia and New South Wales. He is seen in **Plate 56** wearing a light suit. In 1910 he was engaged by Messrs Ehrlich & Co and first went to the Blackwater Mine in New Zealand before being moved to a gold mine near Pilgrim's Rest in South Africa. He then reported on the Pigs Peak Gold Mine in Swaziland, subsequently operating it until its exhaustion in 1917. From 1917-1920 he operated the Mount Morgan Gold Mine near Barberton in South Africa.

He returned to England in 1920 and was engaged in farming and quarrying in Lincolnshire. During the farming depression of 1927 he left for India to become Mine Superintendent for the Indian Copper Corporation. His manager, in a report dated 13th November, 1928 concluded

I would desire to record my appreciation of the valuable advice and suggestions submitted by our Mine Superintendent Mr Sharpley, whose past experience has been invaluable and who thorough and efficient preparation of the mine has placed us

2 Cornish Post & Mining News, 30 June, 1898.

in a position to regard with confidence the economical extraction of ore in the future.

Sadly, Harold Sharpley was never to see this report as he was killed the following day in an accident.

Captain Temby, as we recorded, died in March 1904 having been in charge of much of the detailed work in practical mining since 1895. In a fine gesture William Thomas recommended that Captain Temby not be replaced until later in the year and that his salary be continued to be paid to his widow for that period. The General Committee thought otherwise and granted the sum of £25 to Captain Temby's widow. Mr Josiah Luke was appointed Mine Foreman. Unfortunately his period as Mine Foreman was short-lived as he died in June 1905 to be replaced by 'Kit' Bennetts, who had joined the mine in 1901, and was to serve as Mine Foreman until the late 1930's. We will meet him again later.

Early in 1906 three more of the staff passed away. Joe Trevarthen aged 72 died after 7 years at King Edward and Thomas Richards, storeman aged 76 after 4 years. They were followed by Capt Cook, then only 49 who had replaced the late Capt Temby as underground foreman. Kit Bennetts replaced Capt Cook and the position of Underground Captain and Foreman was made into one.

Plate 56. Harold Sharpley

Progress and Problems

Mr. J.G. Lawn, an experienced mining engineer[3], joined as Head of Mining and Mine Manager in December 1906. He immediately recommended the installation of machinery to dress the burnt concentrates which would then release the main plant for concentrating ore from the stamps. This consisted of a second Frue vanner and another buddle. **Plate 57**, taken in 1908, shows in the right background this new Frue vanner with, to the left at the back of the mill beyond the Acme table, the three vats of the gold cyanide plant. **Plate 58** is a closer view of the vanner taken at the same time. The additional buddle was constructed outside the mill building to the right of the vats. Production was beginning to increase and it was planned to run the stamps for 10 to 12 hours per day.

Plate 57. Mill – looking south – Frue vanner in the foreground, Acme table on left and second Frue vanner in background towards double doors – 1908

Mr Lawn investigated providing additional dressing water by either recirculating used water or by pumping from underground. He wrote:

Mr Negus (a member of the Mining Committee) assured me that there was plenty of water at our 40-fm. level in the dry season to keep the plant going and our mine foreman is also of the same opinion. There are two schemes for pumping the water either one of which would be possible. The first is to put down two Cornish 5 inch plunger lifts, one at the 40 fathom level and the other either at the adit or at the 10 fathom level and to work them from the mill engine by the use of gearing and rods. This scheme would cost about £200 to instal. The other scheme is to put in an electric pumping arrangement; the generator to be run from the present mill engine. This scheme would cost some £350 and the advantage of it would be that the students would have an example of electric plant on the mine to study. It would also be an advantage to have electric lights on the mine in the winter mornings and evenings. The scheme of returning the water again and again would only be advisable in case we could not get sufficient clean water from underground.

Neither scheme was progressed, possibly because, at that time, Woolf's Shaft at Great Condurrow was being de-watered and the manager of that mine was quite willing to let King Edward Mine have any water which they did not need for their own plant. Great Condurrow had taken over the old South Condurrow stamps and dressing floors, the water from the Woolf's Shaft being pumped to the dams behind the South Condurrow stamps just to the west of the count house. Mr. Lawn also reported:

the students are not working in the mine at all regularly and as the examinations draw near I am afraid the mine work will suffer still more (a situation which has not markedly altered in 90 years!)

By mid-1907 the additions to the dressing floors had been completed, the new winding engine had been ordered from Holmans and work was advanced on the hoist foundations. Underground the 10 east and the 20 east on William's Lode were being driven by hand, the 30 east was being driven by rock drill. Pipes were being laid into the 40 east preparatory to machine driving as it was too hard for hand work. The 10-fm east was running at about 60 lbs. of black tin per ton, the 20-fm east was also fairly good but the 30 east was poor.

Two students, during the 1907 August holidays, had taken the 10-fm end to drive at £5 per fathom. They drove 4 feet 3 inches in a fortnight, or only about 4 inches per day! The rock drill miners were being paid £5 per month and the other miners £4 15s. A hand stope above the 30-fm level on William's Lode was also being worked – operations on the Flat Lode had by this time nearly ceased.

A considerable amount of maintenance was needed in the shafts and the main shaft was lined above adit preparatory to running both cages with the new winder. The North Shaft had been started as a provision in case Wheal Grenville stopped work, though what use it would have been is not clear, was beginning to cave from surface and was repaired. Mr. A.B. Climas[4] resigned on appointment as Manager of Wheal Sisters and Mr. Hopley was promoted to his position as Head of Survey.

The new boiler was not performing well and Mr. Lawn recommended the installation of a feed water heater using waste heat. None of the engines were condensing and the boiler was fed with cold water. The draught was also very bad, and at times the plant had to stop for want of steam. Mr. E.C. Murdoch, a Consulting Engineer, later reported on the system:

3 After graduating from the Royal School of Mines in London in 1888, he became chief surveyor with the Barrow Haematite Company. Following a period lecturing at RSM he went to South Africa as Principal and Professor of Mining at the new School of Mines at Kimberley. In 1903 he became assistant consulting engineer to the Johannesburg Consolidated Investment Co. (J.C.I)

4 A B Climas, who had studied under Willie Thomas, went on to become the manager of Botallack in 1910.

There is no doubt that especially with certain winds the draught in the boiler is very bad and not sufficient properly to consume the fuel nor to raise steam to work the machinery. In my opinion the arrangement of some of the machinery has been very badly laid out. It must have been obvious several years ago that a new boiler and winding engine would be required and there should have been sufficient foresight to have left room for the former, where the air compressor now is and it would have been in a convenient position to serve all engines. Then, for no strong reason so far as I could ascertain, the boiler was put in with the wrong end next to the chimney[5] and there is no doubt in my mind this is the principle cause of the bad draught.

A forced draught system was installed to overcome the problem.

In the mill close assaying had shown that the plant was only recovering some 65% of the tin. Most of the losses were in the slimes and the installation of a slime plant consisting of a dozen automatic frames and a round table was recommended. Most of the ore milled was coming from development and production was running between 80-100 tons per month at a grade of 25lbs. of black tin per ton. Milling proceeded without interruption due to any lack of water as Condurrow were not dressing. The mill was rented for two weeks during the summer of 1908 to test ore from the Boswin Mine (Carnkie). Some test milling was also done for Tresavean (Lanner) and Killham Mine (St. Neot). This kept the mill busy during the period when underground production fell off due to the student vacation. Production in 1908 was 971 tons of ore containing 24.5 lbs. of black tin ore per ton. This compared favourably with the

Plate 58. Mill – Frue vanner and feed strip at bottom of mill – 1908

Plate 59. Holman winder at Castle-an-Dinas – 1945

1907 figure of 954 tons at 18.7 lbs. per ton. The improvement in grade was attributed to the fact that operations were confined to William's Lode.

The new winding engine went into service in February 1908. This was a 10-ins by 15-ins coupled geared twin drum steam hoist supplied new by Holman Bros.. There are no known photographs of this winder in its house at King Edward. Later, in 1942, the winder was removed to the Castle-an-Dinas wolfram mine[6] where it worked until 1957. George

5 A Cornish boiler has a single large fire-tube passing down the centre of the boiler from the fire box. The hot gases exiting the fire tube are passed back at a lower level towards the firebox end of the boiler along firebrick flues. Finally the gases enter another flue that lies directly under the boiler and they pass along under the boiler before exhausting into the chimney. These flues serve to heat the lower, cooler parts of the boiler water. At KEM the chimney was at the firebox end so either the final return flue was omitted or a third flue was put in to return the gases to the chimney.

6 See "Castle-an-Dinas 1916-1957, Cornwall's Premier Wolfram Mine" by Tony Brooks. Cornish Hillside Publications, 2001.

Ellis, a well known photographer, recorded it at 'Castle' on 26th June 1945. **Plate 59**. The winder was subsequently rescued and put on display at the Poldark visitor attraction. At the time of writing (2001) the winder is back at King Edward awaiting re-erection on its original foundations.

Plate 60 shows the newly erected winder house just to the right of the headframe. Just visible between a pair of tramway legs is the compressor house with the larger mill engine house to the right. Far right is the tall building that housed the Californian stamps. With modifications to the boiler, steam pressure could be maintained at 70 lbs per square inch. May 1908 saw the removal of the old beam winding engine by Holmans. Presumably it was scrapped as there was little market for an old steam whim. During the same month, Josiah Jewell[7] fell down on the incline used for raising dump on the surface near the shaft. This is the incline that can be seen in Plate 51. According to the accident report: He was only bruised but being 71 years of age it was something of a shock.

Whilst the centre of the School's activities was in Camborne many meetings were held on the mine. To attend these meetings the School's management

Plate 60. Engine Shaft – horizontal winder, mill engine and stamps houses – about 1909

Chapel Street, Camborne,
Folio 405
Dect 31st 1909.
Messrs Camborne Mining School
Dr. to The Camborne Posting Company, Ltd.
(ESTABLISHED 1891)
Cab Proprietors and Postmasters.
Telephone No. 55. Cheques and P.O.O.'s payable to the Camborne Posting Co. Secretary: W. Rowe.

1909					
Oct 7	To Carriage to & from King Edward Mine J.J.B.		2	6	
Dect 11	" do " " Do. 1½ hrs. Mr Kn.		3	.	
11	" do " Trevu . C.V.J.		1	0	
			6	6	

Received Feb 1/1910
Camborne Posting Co Ltd
Per W Borlase

CAMBORNE MINING SCHOOL
Passed for Payment

Figure 9. Camborne Posting Company Invoice – 1909

would hire transport from the local cab proprietors, as most people did not have their own horse or carriage. **Figure 9** is an invoice for transport to and from King Edward for J.J. Beringer and T Knowles, the Secretary. Horses were still the main haulage system on Cornwall's roads with bulk materials being delivered to the nearest railway station before collection by horse and cart. Coal for King Edward came by rail from the Radstock Collieries near Bath, **Figure 10**. Most other materials and equipment required on the mine came from local suppliers and manufacturers.

Plate 61, taken from Gould's engine house of Wheal Grenville, about 1912, shows the site in its final form. Note the calciner with the flue exiting the building at high level before dropping down and turning the corner to the stack. The dark painted lean-to structure immediately to the right of the calciner was put up over the extra buddle mentioned earlier in this chapter. The building was, as we shall see later, converted into a laboratory. The roof of the new winding engine house can just be made out to

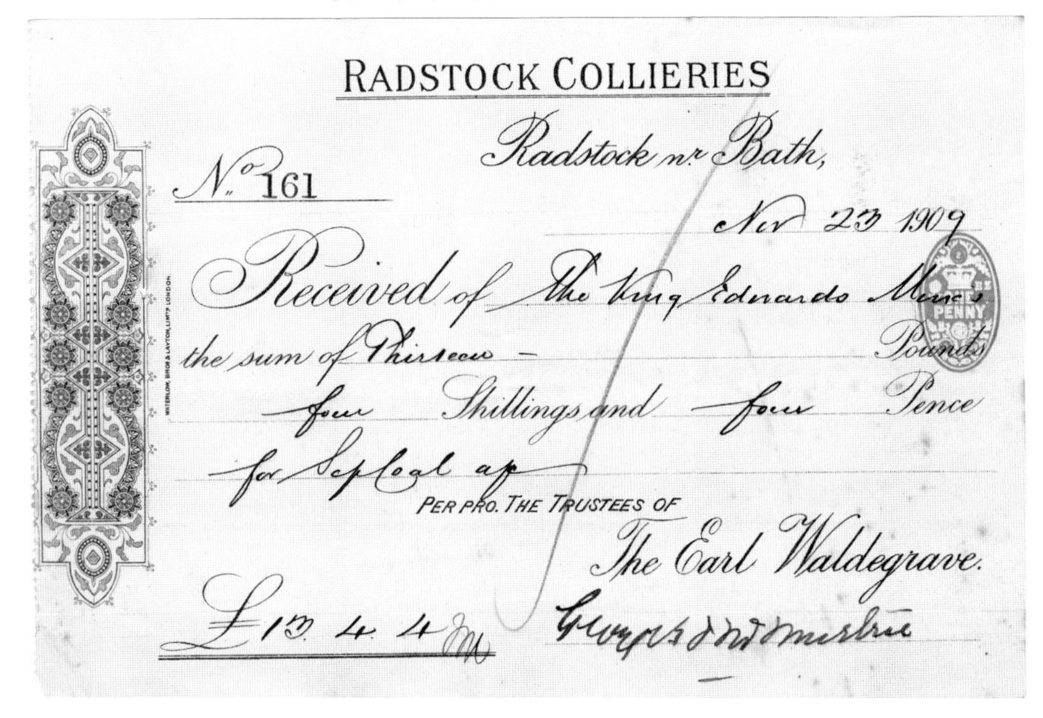
RADSTOCK COLLIERIES
Radstock nr Bath,
No. 161
Nov 23 1909
Received of The King Edwards Mine
the sum of Thirteen - Pounds
four Shillings and four Pence
for Sep Coal a/c
PER PRO. THE TRUSTEES OF
The Earl Waldegrave.
£13. 4. 4

Figure 10. Radstock Collieries receipt – 1909

7 Josiah Jewell was born in Breage and, in the 1881 census, his occupation was recorded as being a tin dresser. He was then living with his wife and two children at 12 Pendarves Row, Troon.

the right of the cloud of steam and in front of the derelict whim house which seems to have lost its bob platform supports.

No further structures of significance were added after this time and all of the buildings, except the horizontal winder engine house and the compressor house, survive to this day. **Figure 11**.

Plate 61. Surface view looking north from Gould's engine house – about 1912

Figure 11. Plan of Layout of Surface Buildings – 1912

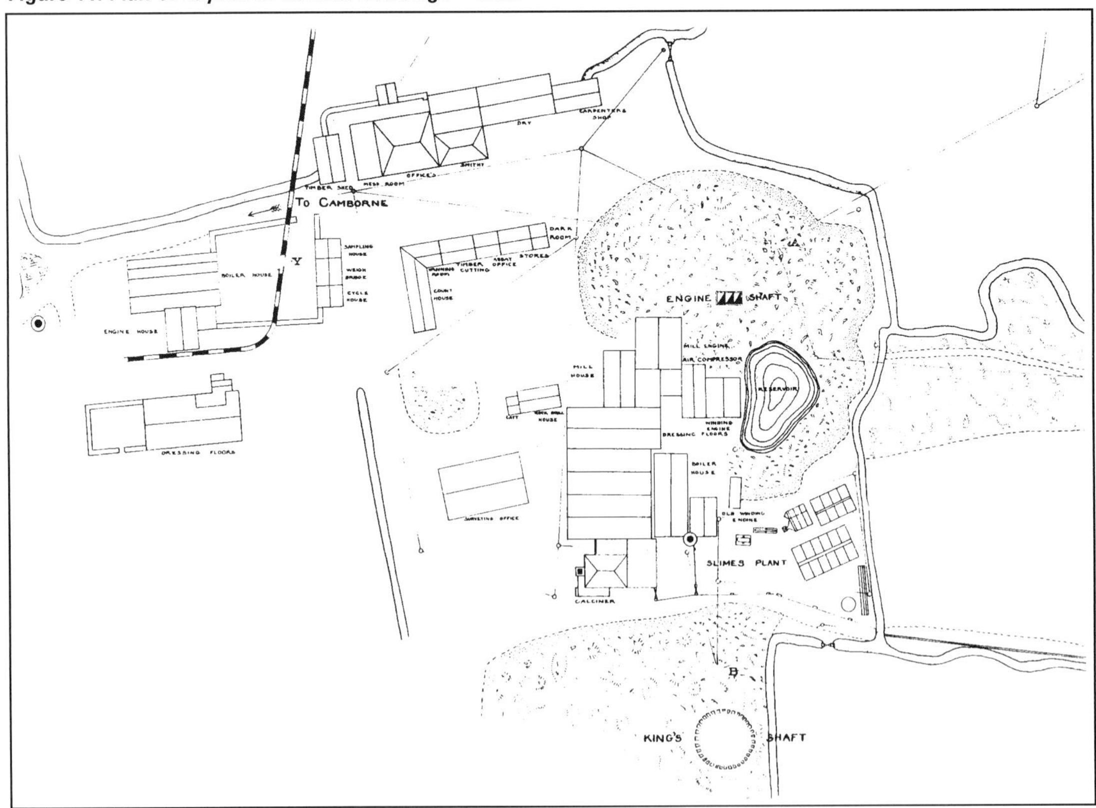

Chapter Six
Decline

A few years ago a number of very faded brown photographs were found in the School's archive. When re-photographed using modern enhancement techniques, they turned out to be a set of pictures of King Edward taken by a student in 1908. Whilst the quality is still not good they give us a glimpse of life at King Edward through a student's eyes. We are not certain who the photographer was but we believe it was Ralph Henry Tredgold, who appears in several of the photographs.

At this time the Great Condurrow mine, which lies on top of the hill immediately to the north of King Edward, was being re-opened. The Great Condurrow company had also taken over the old South Condurrow stamps and dressing floors. An inclined tramway was built to link the underground mine with the dressing floors. **Plate 62**, is the only photograph that we have found that shows this tramway. This unique view of King Edward from the north was taken looking down the tramway. Water, pumped up from underground at Great Condurrow, flowed down the launder seen here and into the dams just to the west of the South Condurrow stamps. It was this source of water that kept King Edward operational until 1914. That King Edward is working is evidenced by the exhaust from the winder and the open door to the stamps at the end of the elevated tramway from Engine Shaft. Smoke is also coming from the calciner stack. Across the valley, Grenville New Stamps can be seen.

Plate 62. Tramway from Woolf's shaft of Great Condurrow to the South Condurrow stamps – 1908

The long building in front of the stamps engine house was the vanner house.

Surveying was an important subject and we have what appears to be a survey group, **Plate 63**, posing somewhat self-consciously by the survey office.[1] Note the team has a complete set of surveying equipment including a theodolite, ranging rods, a tape, and recording books. Note also the clothes that the students are wearing, smart suits, white collars, waistcoats with watch chains etc. – very different from the 'casual look' we see today.

Going underground was what it was all about and in **Plate 64** we see the Mine Foreman 'Kit' Bennetts in the centre, with Hayes to his right and Tredgold to his left. They are standing in front of the headframe.

Plate 63. A Survey Class - 1908
Back L-R: S Y Wong, A Beringer, M C James.
Front L-R: W Beringer, R H Tredgold.

Plate 64. Engine Shaft headframe – L-R: Hayes, Kit Bennetts (Mine Foreman), R H Tredgold. Engine Shaft – 1908

The practical mining staff feature in **Plate 65**, taken in almost exactly the same position as Plate 64. Unfortunately none are identified. To the right is a concrete foundation on top of which is bolted a metal plate supporting one of the headframe legs. During excavations to stabilise Engine Shaft in 1997, a plate identical to the one in this photograph was recovered.

Finally the students are about to go underground. Three friends, in clean underground kit, stand by the entrance to the shaft, **Plate 66**. Possibly on the same day, another group wait to go underground **Plate 67**. Kit Bennetts can just be seen behind the group.

In **Plate 68** we see our happy group on the 30-fm. level. We do not know what type of photographic

1 W.Beringer died in 1918 and M C James in 1935. Tredgold is recorded, in 1966, to have been living in South Africa.

Plate 65. **Group of miners on surface at Engine Shaft – 1908**

Plate 66. "Going underground" – Dalzeil, R H Tredgold and Brydon – 1908

equipment was used for this last photograph. Flash photography, taken for granted today, would have been unusual for the amateur in 1908.

Mr. Lawn, who had been Mine Manager since 1906, left in July 1909[2]. A temporary replacement was sought in view of the impending amalgamation of the mining schools in Cornwall. At that time there were mining schools in Penzance, Camborne and Redruth. Camborne was by far the largest and most progressive of the three. The schools were shortly to be amalgamated at Camborne under the title – the Camborne School of Metalliferous Mining. In the

2 On leaving CSM in 1909 he returned to South Africa as Principal and Professor at the South African School of Mines for a year before rejoining Johannesburg Consolidated Investment Company as consulting engineer. During the 1914-18 war he worked with the Explosives Department of the Ministry of Munitions and then, in 1919 went back as managing director of J.C.I. before becoming, in 1924, director and consulting engineer in England. He died in 1952.

Plate 67. Group of students about to go underground. W Berringer (third from right), S Y Wong (front) – 1908

Plate 68. Student group on the 30-fm level. Kit Bennetts front right – 1906

interim a Mr. H. Jenkins (late of Wheal Jane), who was teaching part-time at the Redruth School of Mines was appointed temporary Manager. Mr. Jenkins's first action was to increase production and to mill on two shifts. Some additional labour was employed but his efforts were apparently frustrated by poor student attendance. Relations between Mr. Jenkins and the students went from bad to worse and, in March 1910, Mr. Jenkins submitted his resignation. William Thomas, who by now had left Botallack, was engaged on a temporary basis. He was not impressed with the results achieved through increased tonnage and longer milling hours and reverted to the old system of one shift working. This resulted in an increase in the grade of ore milled and returns were largely unaffected.

'Willie' Thomas's temporary engagement terminated in July 1910 on the appointment of Mr. W. Fischer-Wilkinson[3] a consulting mining engineer of some eminence, as Principal of the newly-amalgamated School of Mines. It is sad to note that after 20 years devoted service J.J. Beringer was not even made Vice-Principal.

One of Mr. Fischer-Wilkinson's first actions was to put the mill back on to two shift milling in an attempt to reduce the annual loss of about £1000 that was incurred at the mine. What the staff at the mine thought of yet another reversal of policy is not recorded.

Underground there was the perennial problem of decay of the timber bulkheads over the levels. Fischer-Wilkinson, who certainly was not short of ideas, experimented with a reinforced concrete bulkhead[4]. This was tried in a section of the 30-fm level, in place of the conventional timber. It was an experiment that seemed to work. Another innovation was the installation of a hooter which gave 2 blasts at 8.30 and some said was used to remind students to get up. It must have been some hooter to have woken a student living in Camborne! More likely, considering the relatively late time of 8.30, it was used to signal the start of the working shift.

Decline

Financially the School was struggling and there are frequent recommendations in the Finance Committee minutes during 1911 to reduce expenditure. The main shaft was deteriorating badly and, despite financial restraints, repairs, estimated to cost £500, were started. Mr. Fischer-Wilkinson resigned in February 1912. He had been at cross purposes with the Governors and had been unable to act as he had wished owing to the financial situation. He was replaced as Manager by J.C. Sheppard on an annual salary of £225 (a saving of £575). J.J. Beringer was appointed Principal, and remained so until his death in 1915.

This poem, from the July 1911 edition of the School of Mines Magazine, is a description of the mine at King Edward, the 'Forty' being the bottom level or the 40 fathom level below adit.

Riding down to "Forty" in a narrow skip,
With a lady cousin and a tallow dip
Stood a gallant student, stalwart was his frame,
Slender his companion, never mind her name.

"This is where we labour, all the morning through
Rend the rock asunder, drill and blast and hew;
Learn to drive a level, learn to sink a winze,
Learn to mark a station; and lose it for our sins."

Through the mine they wandered, scrambled up the stope,
Dropped their candles often, often had to grope,
Knocked their heads together, could not find a light,
Stumbled into puddles, wandered out of sight.

Wielded the hammer, swung it like a mace,
"Reggie, I'm so sorry! Was that – was that your face?"
Saw the mighty rock drill eating up the rock.
Planted deep the dynamite and waited for the shock.

Soiled to the surface swiftly they were borne,
Where the sun shines brightly o'er the nodding corn.
Back to lovely Camborne, nestling in the vale,
Down the heights of Beacon, and this concludes my tale.

3 Fischer -Wilkinson was educated at Harrow and Friberg. In 1887 he joined Messrs. Bainbridge, Seymour and Co, obtaining mining experience in Mexico, USA, Africa, Europe and Asia Minor. In 1895 he was appointed consulting engineer to the African Estates Company of Johannesburg and 4 years later joined Consolidated Goldfields. He died at the end of 1942 aged 78

4 Paper published in the Mining Magazine 12th December, 1912.

Plate 69. Hand drilling contest – about 1910

It is interesting to look back at this and wonder whether it was based on fact and if so, who was the young lady?

The magazine also tells of an event at King Edward Mine.

> *The drilling competition at the mine this summer was a happy thought of Mr Shepherd's (Head of Mining). The sight of so many students* (**Plate 69**) *flogging drills into granite blocks in the broiling sun amongst a crowd of admiring miners, bored visitors and sarcastic students was enough to give one food for reflection to last many days.*

By 1912, despite amalgamation, student numbers had fallen to 97, less than half of the 1903 total. On the practical mining side the mine was well utilised. Work underground was now almost totally concentrated on William's Lode. **Figure 13** shows the state of development at the end of 1912.

A student[5] has left us his observations on the underground mine:

5 Alfred Alexander Jones (1910-1913). Report dated June 11th 1913.

The Adit Level – this level is now abandoned both as a means of drainage and for the extraction of ore. In fact there is only one portion that is ever trodden on, and that is between William's Shaft and a winze connecting it (adit) to the 10 fathom level. In this portion it presents the appearance of many years of neglect, so weathered are the walls and the supports.. Here, where it is open for inspection, it is driven in the killas, and the junction of the killas and the granite is seen in this level at William's Shaft.

Descending on a tour of inspection, from the adit to the lower levels, we pass down a winze, known as "Jacob's Ladder". This is a continuation of a steep incline with steps as means of descent, and then a vertical draft (section of the winze).

The 10 Fathom Level – This level is notable by the fact that in many places its walls are of kaolin granite, and in most places it is supported. This is the longest level on William's Lode, being about 700 ft. on that lode in an easterly direction.

Descending to the 20 fathom level – we find the rock much harder. It is this level that the junction of

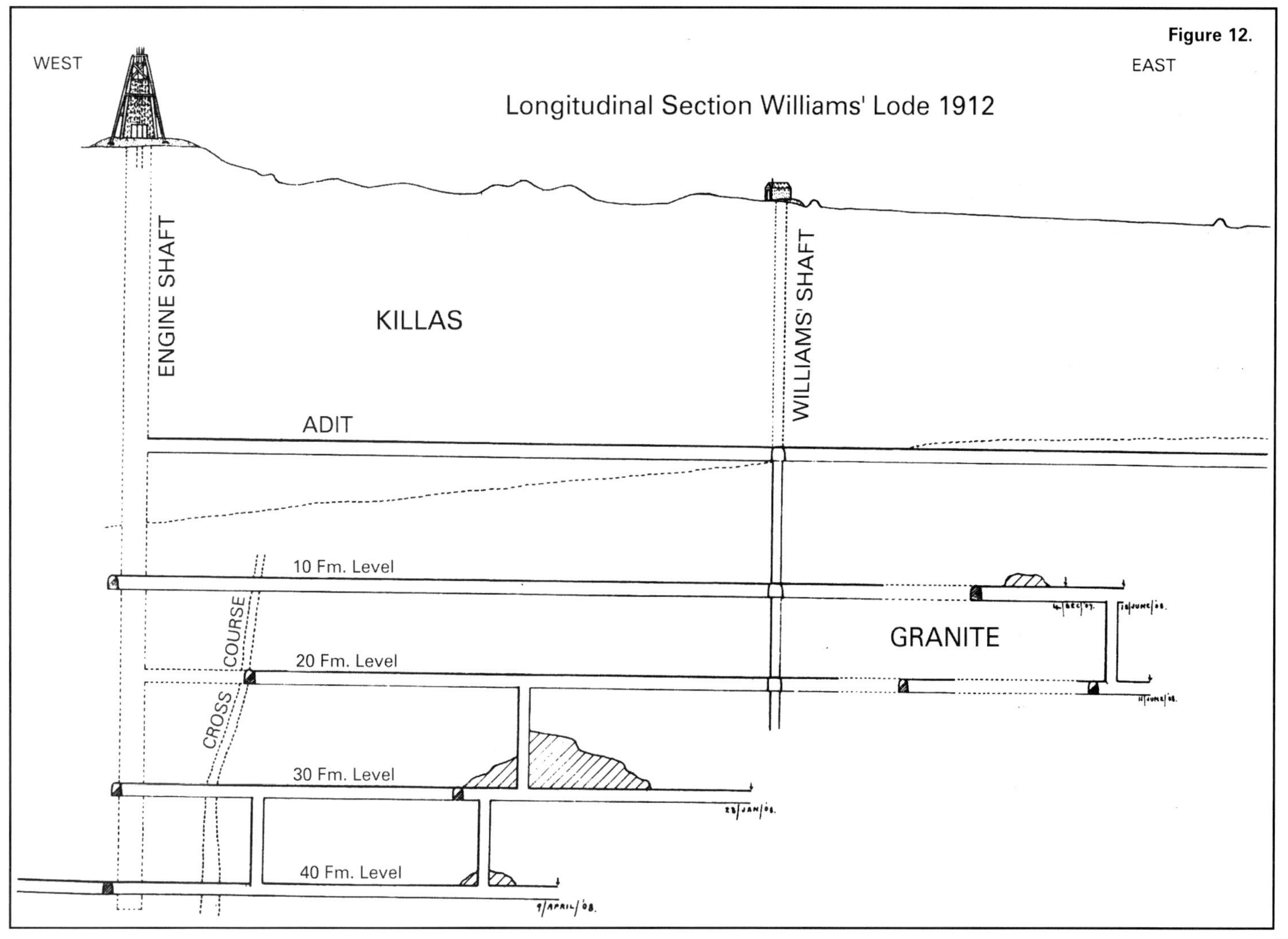

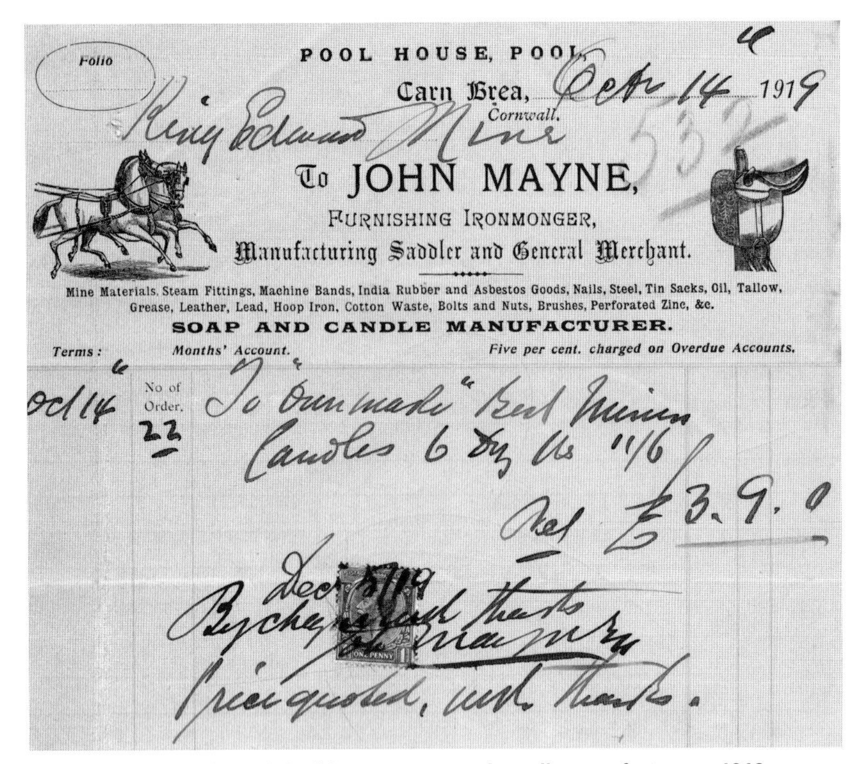
Folio

POOL HOUSE, POOL,
Carn Brea, Octr 14 1919
Cornwall.

King Edward Mine

To JOHN MAYNE,
FURNISHING IRONMONGER,
Manufacturing Saddler and General Merchant.

Mine Materials, Steam Fittings, Machine Bands, India Rubber and Asbestos Goods, Nails, Steel, Tin Sacks, Oil, Tallow, Grease, Leather, Lead, Hoop Iron, Cotton Waste, Bolts and Nuts, Brushes, Perforated Zinc, &c.
SOAP AND CANDLE MANUFACTURER.

Terms: Months' Account. Five per cent. charged on Overdue Accounts.

	No of Order.		
Oct 14	22	To ownmade "Best Miners" Candles 6 doz lbs 11/6	£3. 9. 0

Recd Dec 5 1919 By cheque with thanks

Price quoted, with thanks.

Figure 13. Invoice from John Mayne – soap and candle manufacturer – 1919

William's and Engine Lodes is well seen. Some 300 ft. along the main level we find the level branching into two, the more southerly on Engine Lode and the more northerly on William's Lode. About another 100 feet in an easterly direction a crosscut connects these drifts. The drift on Engine Lode is not now worked, only used as a safe place for dynamite.

At the 30 fathom level – towards the end of the level, we find again the junction of the two above mentioned lodes, but Engine Lode has not been worked. Very near the junction the level becomes 15 ft. wide. Much Native Copper has been found at this depth, with an equal amount of Cuprite, of which quite excellent crystals have been found. Further east we find that an aplite dyke has interfered with the lode values. The ground then becomes softer and in places is highly kaolinised. Going back to the shaft (Engine) at this depth we find a crosscut leading to a drive on the Flat Lode, and we are struck by the size and extend of the stopes (if we haven't been to larger mines)

At the 40 fathom level – on William's Lode there is about 370 ft. of drift, and this level is connected to the 30 fathom level by a rise, about half way in an easterly direction along the level. At this level the Flat Lode has been extensively exploited, and generally speaking has low grade "stuff". On the Flat Lode the 40 fathom level is connected by stopes to the level above, and ascent and descent is made by use of chains. There are other levels below, but their present state of repair is too dangerous for working in.

*Illumination – paraffin wax candles are now chiefly used although acetylene lamps are used by the Manager and Foreman. The candles cost about 4d. per pound, while the calcium carbide is bought at a rate of about 3d. per pound.***Figure 13** *But candles have many advantages, eg. clay will not stick an acetylene lamp to one's hat while climbing up or down a shaft or winze. Also acetylene lamps will not grease holes which may be obstinate which candles will.*

Labour – Seven skilled miners, two trammers and a mine foreman form the underground staff. On surface there are 16 hands all told. Contentment seems to prevail, and all seem to work for the mine's good as well as their own, and in addition the help this working staff gives the students is invaluable

On surface, vanning (Appendix II), panning and timber-cutting were taught. In the second year, the students were divided into sets of four and carried out, with one miner, all the timber repairs required. The application of machine drilling to driving, sinking and stoping was taught. Track-laying, pipe-laying, hand drilling and blasting were also carried out. On surface all sets passed through a course in the blacksmith's shop. Rope capping and splicing, machine drill and winch overhauling, pipe-laying, belt repairs and general engineering repairs were done under the engineer. All students spent a day or so on the winding engine and also worked in the mill. In the third year, the sampling and valuation of William's Lode and a part of the Flat Lode were carried out. Assay plans and sampling reports were prepared. All samples taken were assayed by the students taking them. In the mill a course of mill-testing was carried out, all samples being taken and assayed by the students and the results of the tests recorded. A course of structural engineering was undertaken and each student had to prepare designs of a calciner and house or of a headgear.

In September 1914 the Manager, reviewing the past year, reported that no work had been done on 10 level with the exception of a few repairs. Driving the 20-fm level was advanced very slightly; the work being done intermittently by students. Stoping above the level continued. The average value of the face being worked was about 40 lbs, the cost of stoping here was about 9 shillings per ton. The sinking of William's Shaft from the 30-fm level to the 40-fm level was started in October 1913, and was carried on by one miner with the help of students during most of the year. Hand-drilling was used for the most part, the rock drill being put in for a short time only. The shaft was being carried down 12 feet by 5 feet. The distance sunk being just over 4 fathoms at a cost of sinking of £30 per fathom. Some stoping was done on the 30 stope at William's Shaft during the year mainly for educational purposes. An underhand stope was started in this level for educational work, the ore turned out was of good value running about 30 lbs per ton over a width of 10 feet. Driving the 40 level east on William's Lode was continued, the level was also widened out to receive William's Shaft. The ground opened up during the year and from here to the level above gave about 1,000 tons of ore with an average assay value of 28 lbs. per ton. The cost of driving was £12 per fathom. An underhand stope was worked in the Flat Lode 30-fm level for educational purposes at the beginning of the year; but due to low values it was abandoned. From the assay plans made during the year the available ground was shown as about 6,400 tons with an average value of 17.6 lbs. per ton. Of this amount about 2,000 tons were shown as running about 6 to 10 lbs per ton. The mill ran for 2015 hours out of a possible 2389 and treated 1240 tons giving just over 6 tons of tin concentrate. This was a creditable effort with the mill running for an average of over 40 hours per week

Staff on the mine at that time consisted of:

Underground Miners	5
Timberman	1
trammers	2
Lander	1
Surface Engine man	1
Stoker	1
Smith	1
Striker	1 (boy)
Dryman	1
Mill men	5 (2 boys)
Account House	1 (woman)
Laboratory	1 (boy)
Survey Office	1 (boy)
Carpenter	1
Fitter	1
Mill Foreman	1
Mine Foreman	1
Total	26

As it happened, this was to be the last report on the mine fully working. With the outbreak of the First World War operations at the mine were drastically curtailed. On the 19th of September 1914, all but seven of the employees at the mine were discharged. The mine was reduced to a state of care and maintenance and all underground work ceased. **Plate 70** was probably taken at this time. The man holding the vanning shovel is thought to be W. Warren, the foreman dresser. The mill has a neglected look and the drive belt for the Buss table is missing. On the right is the 'new' round frame that replaced the Acme table.

At the end of the settling pit is a small revolving screen.

Poetry was not only restricted to students since 'Kit' Bennetts at the mine wrote several verses entitled "Miners All" including:

Up miners now, and take your stand
Against this mighty foe,
And teach the Germans, if you can
He'll reap what 'ere he sow......

The aged ones, who stay behind,
Will keep the pumps all right,
And you will find your places dry,
When you've finish with the fight

One cannot imagine the miner of today writing something like this – it was a very different world.

By June 1915, student numbers were at an all time low as most of the students and many of the staff, including Mr Sheppard, had joined the Royal Engineers. The same thing had happened at Redruth which still ran courses as an out-station for Camborne – the number of students at both places was not expected to be more than 25. Rather than release these remaining students for munitions work and only run evening classes, it was proposed:

...only to work certain portions of the mine and avoid the treatment of large quantities of low grade ore. The William's Lode lends it (sic) *to this method of mining. The students with the mine staff would,*

Plate 70. Interior of the Mill – about 1914

for example, break and select ore underground for 2 or 3 months and then the students with the tin dresser would treat this ore for say a fortnight. The hoisting and milling plant would thus be in use for a period every two or three months. This would occupy the students on average 3 days per week, and allow the other 3 days for work at the School, whereas the employment of students as munitions workers would tie them every day and leave only the evenings for instruction.

This arrangement was accepted and production continued.

Things did not go as well as expected, for the shaft was soon found to need further repairs from 40 ft. above adit level to 60 ft. below adit. During the repair period some ore was broken on the 10-fm level and hoisted to surface through William's Shaft using a hand windlass.

Closure

The end of the war, in 1918, signalled a dramatic increase in student numbers and, despite severe financial restrictions, good progress was made in restarting mining. The boiler was checked and repaired by the Cornwall Boiler Company, **Figure 14**, and amongst other materials, wire rope was purchased from John Stephens & Sons of Falmouth, **Figure 15**. By May 1920, five men were employed full-time underground and the staff on the mine totalled 12. 'Kit' Bennetts was still in post as the mine foreman. Unfortunately this revival was to be short-lived. Wheal Grenville had, for some time, been in financial difficulties. The tin crash of 1920 was the final straw. Grenville was abandoned in the middle of 1920 and the mine was left to flood.

This left King Edward very vulnerable. Enquiries were made about the existence and condition of the Grenville adit, but no one seemed to know much about it. At a special Mining Committee meeting on 20 August 1920, it was reported that:

It seemed probable that it was the same as was figured on an old Home Office plan as the Deep Adit about 50 fms and opening near the Penponds viaduct. If this was clear the work at King Edward could be continued as usual. If this adit were choked the water would probably rise so high in King Edward as to make it impossible to work there.

The water was rising in Grenville at about 18 inches per day and was by then up to the 100 fathom level. It was mentioned that:

In case the Deep Adit of Grenville is choked it might be worthwhile to reconnoitre Marshall's Shaft

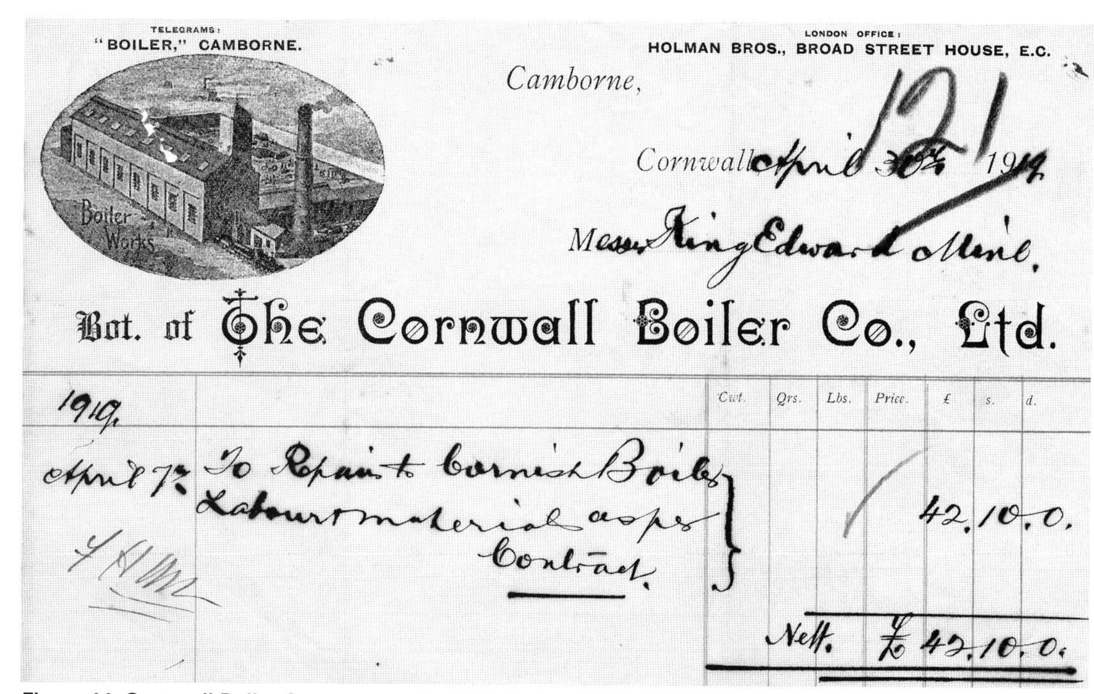
TELEGRAMS: "BOILER," CAMBORNE.

LONDON OFFICE: HOLMAN BROS., BROAD STREET HOUSE, E.C.

Camborne, Cornwall April 30th 1919

121

Messrs King Edward Mine.

Bot. of The Cornwall Boiler Co., Ltd.

		Cwt.	Qrs.	Lbs.	Price.	£	s.	d.
1919.								
April 7th	To Repair to Cornish Boiler Labour material as per Contract.					42.10.0.		
	Nett					£42.10.0.		

Figure 14. Cornwall Boiler Company invoice – 1919

which it is understood could easily be isolated from Grenville. It might also at the same time be worthwhile investigating the possibilities of Great Condurrow.

The attempt to re-open Great Condurrow had ended in failure in 1914.

By December 15th 1920 the lower levels of King Edward were flooded and the water had risen to the 10 fathom level. Regrettably King Edward's deep adit level was choked and the water continued to rise. By February 1921:

The water had now risen beyond the shallow adit which was choked, and which the students with the help of the miners we are now engaged in clearing.

Whilst clearing the shallow adit kept everyone occupied, this was a temporary expedient and offered no permanent solution as a single adit was hardly sufficient for the School's work.

Great Condurrow, which was known to be dry to 50 fathoms was again considered. In one of the shafts it had been ascertained that there were several old workings above the 50-fm level suitable for the School's purpose. The Principal, Mr A. Richardson, said:

He proposed to drop ladders during the holidays in this shaft (sic) referred to, and to explore the possibilities of the terrain.

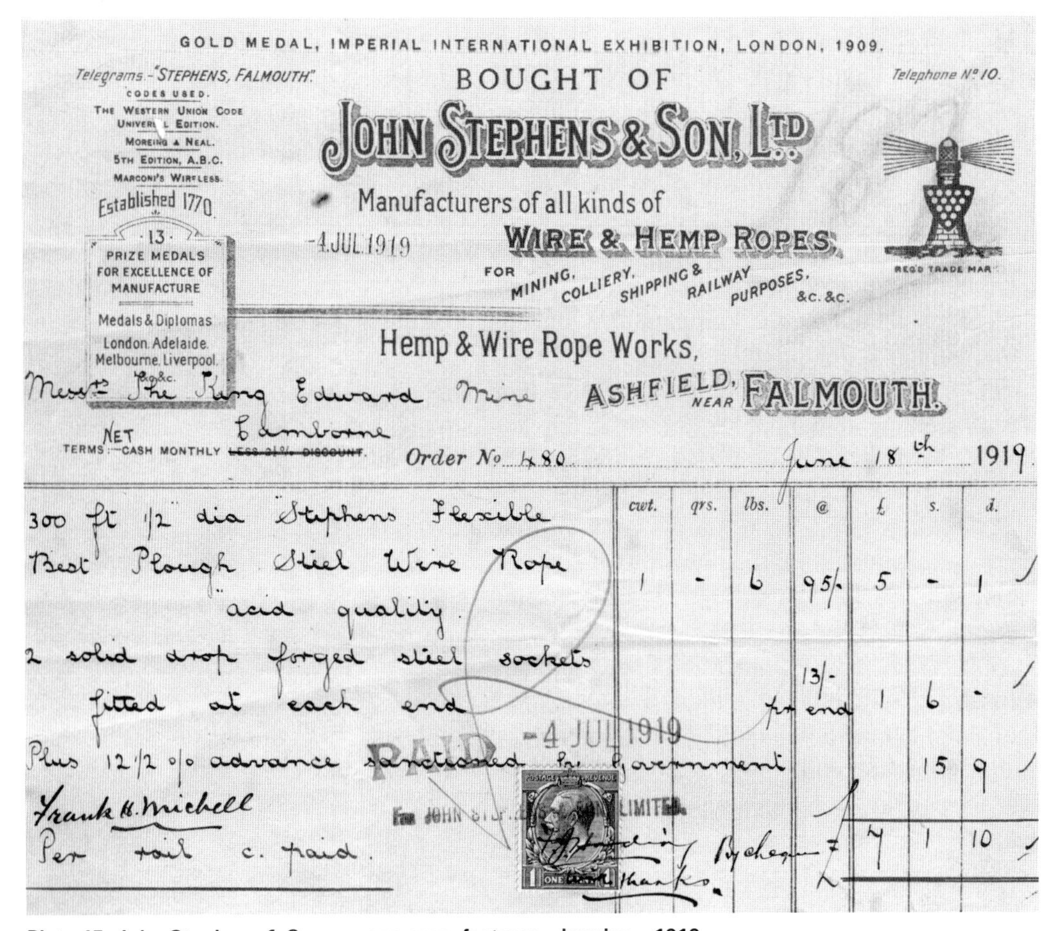

GOLD MEDAL, IMPERIAL INTERNATIONAL EXHIBITION, LONDON, 1909.

Telegrams.-"STEPHENS, FALMOUTH."
CODES USED.
THE WESTERN UNION CODE UNIVERSAL EDITION.
MOREING & NEAL.
5TH EDITION, A.B.C.
MARCONI'S WIRELESS.

Telephone Nº 10.

BOUGHT OF

JOHN STEPHENS & SON, LTD

Established 1770

13 PRIZE MEDALS FOR EXCELLENCE OF MANUFACTURE
Medals & Diplomas
London. Adelaide. Melbourne. Liverpool. &c. &c.

Manufacturers of all kinds of
WIRE & HEMP ROPES,
FOR MINING, COLLIERY, SHIPPING & RAILWAY PURPOSES, &c. &c.

REGD TRADE MARK

-4 JUL 1919

Hemp & Wire Rope Works,
ASHFIELD, NEAR FALMOUTH.

Messrs The King Edward Mine
Camborne

NET
TERMS:—CASH MONTHLY ~~LESS 2½% DISCOUNT~~

Order No 480 June 18th 1919

	cwt.	qrs.	lbs.	@	£	s.	d.
300 ft 1/2 dia Stephens Flexible Best Plough Steel Wire Rope "acid quality".	1	-	6	95/-	5	-	1
2 solid drop forged steel sockets fitted at each end				13/- per end	1	6	-
Plus 12 1/2 % advance sanctioned by Government						15	9
					7	1	10

Frank H. Michell
Per rail c. paid.

PAID -4 JUL 1919
For JOHN STEPHENS & SON LIMITED.
By cheque
Thanks.

Plate 15. John Stephens & Sons – rope manufacturer – invoice – 1919

Chapter Seven

Near Extinction and New Hope 1921-1939

The closure of Wheal Grenville and King Edward in 1920 marked the end, probably for ever, of mining along the Great Flat Lode. The extensive Basset Mines Ltd., which had worked the Flat Lode to the east of the Red River, had closed in 1918. The constant roar of the stamp batteries, so long a familiar sound, were silenced. The plant and machinery was sold off, the workers went elsewhere, and the valley slowly reverted to nature, with just a few ruins and dumps to show that there was once a big industry here. But not everything was sold off or ruined – King Edward, outwardly at least, remained much the same.

In the post-war period, the School was also going through difficult times and could not afford a full-time principal. Mr A. Richardson, who had only joined after the war, left and the post of part-time Principal was taken by R.A. Thomas, who also continued to be the manager of Dolcoath. The old Dolcoath mine had closed in January 1921, but the company, under his management were shortly to embark on a major shaft sinking project in an attempt to find new orebodies to the north of the old mine. This was not successful and the mine closed in 1929.

There are few records of what happened at King Edward in the 1920's. We know that Great Condurrow was opened up via two small shafts down to about 100 ft from surface, and that a level was established in the old workings at this horizon. These workings were on Landower Lode that lies about 150 ft. north of the Great Condurrow Main Lode. There was no money for a hoist or even a compressor, and compared with the old King Edward mine it was tiny. However mining education was gradually changing and there was no longer the same need for a full working mine. There was plenty of work to be done just opening up the old workings and making them safe and accessible. **Plate 71** was taken about 1921 at the start of the re-opening.

Plate 71. Opening up Great Condurrow – 1921

The surface area of King Edward must be the most surveyed place on earth. Surveying at King Edward has always been a major feature of CSM courses and this poem, printed in 1923, sums up brilliantly the feelings of many a student – of any period!

I was seated one day at King Edward
I was weary and sore distressed,
For one half of my data was borrowed,
And the other half it was guessed.

'Twas in vain I attempted to plot it,
My plan was a horrible sight;
It had ink smudged all over the corners,
And n'er a co-ordinate right.

And the buildings were mostly lopsided,
For not one of the walls were straight;
Then the hedges were sized like a roadway,
And only a few had a gate.

And I swore as I sat in the office,
Surveyor I'll never become;
I'd soon break stones at Pentonville,
Than toil as a surveyor, like some.

On surface at King Edward it was a case of business as usual. **Plate 72** was taken in 1925 and was taken from close to the headframe, On the right is the miners' dry with the blacksmiths' shop in the centre and the count house beyond. These were the original South Condurrow buildings and date from the 1870's. To the left is the vanning room/workshop block. The little underground operation at Great Condurrow did not produce ore and the mill was now used more as a full-size laboratory.

The Town and Country News magazine carried an article in December 1931 on the underground mine. **Plate 73**, is from this article and shows a group of students hard at work hand-drilling blastholes on the 100-ft level at the western end of the mine. **Plate 74**,

Plate 72. L-R Workshops, Count house, Blacksmith's shop, Miners' dry – 1925

Plate 73. Underground at Great Condurrow.
Western end – 1931/32

Plate 74. Loading a tram – 1931/32

Plate 75. Headframe and buildings at Great Condurrow – 1948

taken at the same time, shows a team of men clearing broken rock from some old stulls.

R.A. Thomas retired in 1933. The new Principal was Major Standish Ball, a mining engineer with international experience[1]. The School and the mine had been starved of capital for years, and one of Major Ball's first tasks was to organise a world-wide appeal for funds. Perhaps in sympathy – recognising that its time was done, on November 31st 1934, the old headgear on Engine Shaft collapsed.

The appeal enabled a great deal of vital work to be carried out. At Great Condurrow, an electrically driven

1 Trained at the South African School of Mines and Technology, now the Witwatersrand University followed by an MSc. in ore-dressing at McGill University in Canada. From 1911 to 1915 he was with the Rand Mines in South Africa. During the First World War he served with the Royal Engineers, being awarded the OBE. After the war he went to the Francois Cementation Company before going to the Trinidad oilfield. After a period in Canada he returned to South Africa as chief surveyor and acting manager of the Sub Nigel gold mine. He left this work to return to oil, this time in Venezuela as resident managing director.

Plate 76. Bar-mounted drifter, Great Condurrow – 1936

compressor was installed. A new 22 ft. high steel head-gear made by Head Wrightsons, tracks, loading gear and electric hoist were erected around the East or hoisting shaft. **Plate 75**, taken some years later, shows the completed works around East Shaft. On the right is the building housing the compressor with its tank for cooling water. Beyond is the fan house with its inclined drift into East Shaft. To the right of the headframe is a short incline. This incline was designed so that hoisted ore could be dumped into a wagon on surface which could then be hauled up the incline to be tipped into a lorry for transport to the mill.

Unfortunately the underground mine produced hardly any ore so the facility was seldom used. To the left of the headframe is the winder house with a small attached workshop.

Underground East Shaft was equipped with the necessary timbering and station equipment. Machines could now be used for drilling underground and the workings were beginning to look like a modern mine. **Plate 76** illustrates the use of a bar mounted drifter. The term drifter refers to the machine's use for driving tunnels or drifting. **Plate 77**, taken in the western end of the mine, shows two men using a stoper. This machine was developed specifically to drill steeply inclined holes. The rockdrill has a pneumatic leg which pushes the machine upwards as the hole is drilled.

Plate 77. Stoper, western end Great Condurrow – 1936

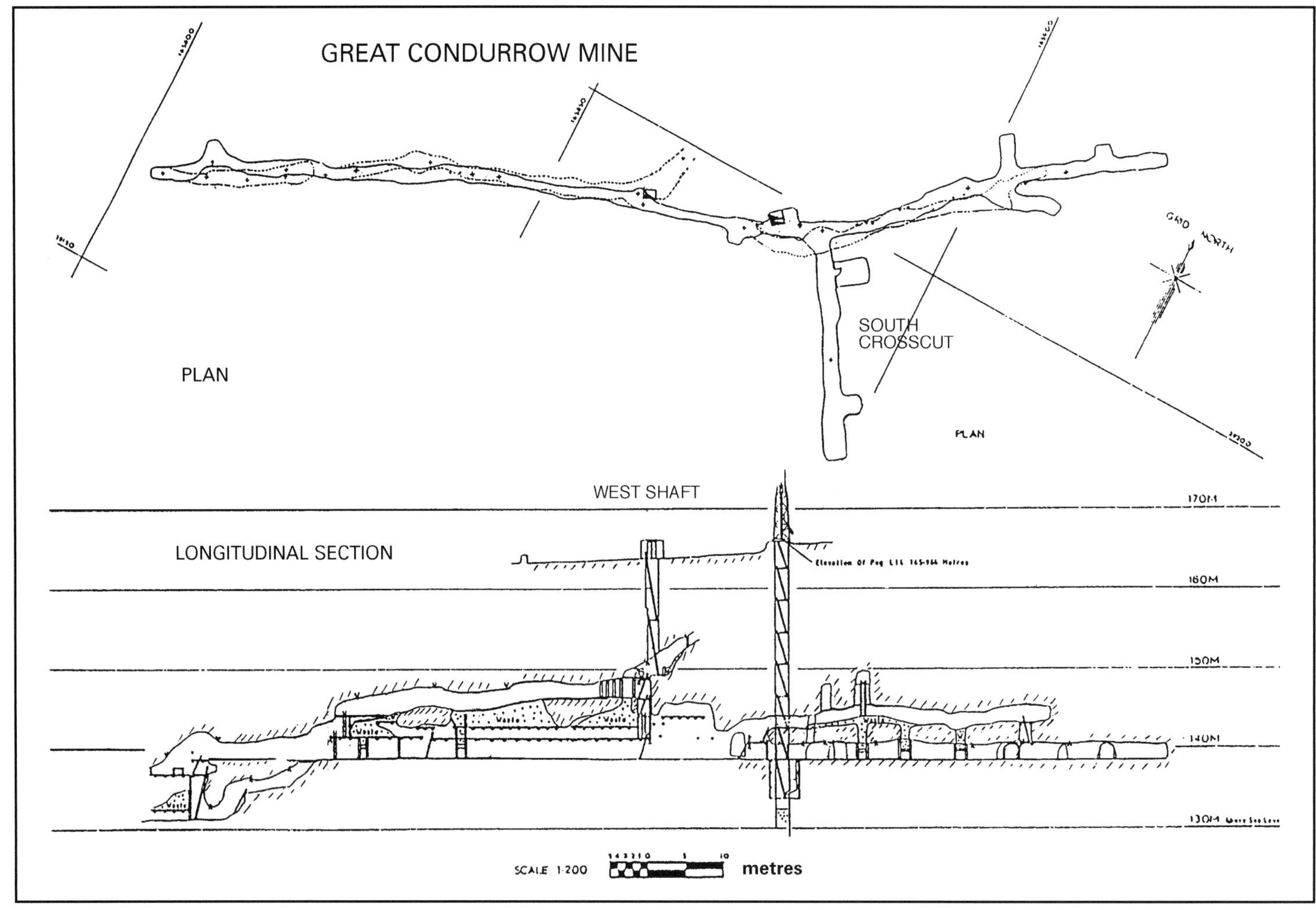

Figure 16. Longitudinal Section – Great Condurrow Mine – 1990

This early machine had no rotation so the miner had to swing the machine to and fro to create a rotation movement. The platform that the instructor is standing on looks very insecure.

A cross-cut was started from the bottom of the shaft, mainly for instructional purposes but also in the hope of opening up a second lode. The underground mine today (2001) consists of two sections: the old 19th Century workings that were re-opened up until 1936; and the new working to the east and south of East Shaft that have been developed since that time and more properly represents the way that a Cornish mine would have looked like in the middle of the 20th Century. **Figure 16**.

The mine was used extensively for underground surveying **Plate 78**. The man standing on the platform is surveying steeply down to the haulage below using a theodolite with a side telescope. These are used when the sight is so steep that the view from the central telescope is obscured by the base of the theodolite. The man below, with his back to the camera, is illuminating the target which is suspended

Plate 78. Surveying underground at Great Condurrow – 1936

from the stull jammed across the excavation. Note the near vertical dip of the lode and the stull timbering creating platforms and head-cover.

At King Edward a mains water supply was connected to the mill. This replaced the system of leats and ponds with which the mine had struggled from the start. An incline rail track was erected to carry the ore up to the stamps (prior to this the old hoist and headframe were used to elevate ore to the stamps), and the dry was completely remodelled.

The mill buildings were extensively renovated and James sand and slime tables were added to the mill **Plate 79**. The slime table is on the left and the sand table is at the back far right. In the centre is the original Frue vanner. In the foreground is the convex buddle.

The old buddle room, which was an extension built onto the south side of the mill alongside the calciner, was converted into an ore dressing laboratory equipped with a jig, ball mill, magnetic separator, flotation plant,

Plate 79. Interior of Mill – 1937

small James shaking table, etc.. In **Plate 80**, the James table can be seen mounted on the table on the right and the jig at the back in front of the electrical panel. Both of these pieces of equipment have survived to the present day.

The rockdrill shop was updated. What was then the latest equipment was installed for instruction in the design, use, care and maintenance of rock drills. New machines were added: drifters, stopers and hand hammers, as well as modern lubricating and dust allaying devices. The shop was equipped with a two stage compressor, oil furnace, drill sharpening machine and hot milling machine. **Plate 81**.

Kit Bennetts must have been pleased to have seen some new money pumped into his beloved mine. Sadly he died, aged 67, on the 19th September 1939, at Redruth Hospital. Born at Ramsgate, just outside Camborne, he worked as a lad in the local mines:

Plate 80. Mill Test Laboratory – 1937

walking great distances to work and climbing many hundreds of feet of ladderway. Later he spent a number of years in Colorado, prospecting and tributing. He returned to this country in 1902 and started underground at King Edward.

He was the last of Willie Thomas's old guard from the early years of the century. King Edward owes much to the loyalty of its mine foremen.

Plate 81. Rock Drill and Sharpening Shop – 1937

Chapter Eight
Marking Time 1940-1969

The Principal, Major H Standish Ball, died in 1941. He was not immediately replaced and the work of running the School fell to the Vice-Principal, George Whitworth. Colonel George Whitworth, as he was often known, was to be awarded the OBE in recognition of his work as Commanding Officer of the Camborne Battalion Home Guard. A new excuse for being late for early morning lectures was: *"Please sir, I was on duty all night."*

As during the First World War, student numbers fell drastically and by the end of 1942 there were only about 40 students in the School. This effected the amount of work at the mine and there was no money for new developments. Only 12 students graduated in 1944 and a shortened course was initiated. A past student was killed when the plane in which he was travelling home from Lisbon was shot down. This was the same plane in which Leslie Howard, the famous actor, met his death. The end of the war saw a dramatic increase in numbers and in 1946 George Whitworth was appointed Principal.

The underground section at Great Condurrow continued to be used for practical exercises and for some small scale tunnelling. **Plate 82** – drilling an up-hole about 1946. The student is using a modern 'stoper', the feed being controlled by a twist grip which he is holding in his left hand. Note the oil bottle

Plate 82. Drilling an up-hole Great Condurrow – about 1946

in the air line on the ground behind the driller, this was designed to provide continuous lubrication in the air feeding the machine. The wagon in the background appears in photographs over the next 30 years and was finally 'retired' from the mine in the early 1980's. It has been preserved above the stamps at King Edward.

The skip in East Shaft could be used both for hoisting rock to surface and for lowering equipment and materials. **Plate 83** shows a group, in 1946, unloading materials from the skip at the main level at East Shaft. The ladder, in the centre of the photograph, runs up the shaft alongside the hoisting compartment and serves as a second access. Just round the corner from the shaft, a group are loading a wagon in the South Crosscut, **Plate 84**. The man in the foreground is using a whirling hygrometer – an instrument that measures wet and the dry bulb temperatures from which the relative humidity of the air can be calculated.

Surveying continued to be an important part of mining education. **Plate 85** shows a group surveying on the main level looking east. East Shaft is just

Plate 83. Unloading the skip Great Condurrow - David Ferguson with hygrometer and Gordon Howard on right, about 1948

Plate 84. Loading a wagon in the South Crosscut, Great Condurrow – about 1948

behind the photographer and the tracks to the South Crosscut can be seen turning out to the right. They are using one of the latest microptic theodolite – mounted on a very new tripod. Note the variety of hard hats worn – from the traditional 'felt' hat in the foreground to the ultra-modern plastic hat worn by the man in the white overalls. Note also the carbide lamps in this and the previous plate. In the background another man is pretending to be pulling a chute.

Holman Bros. Ltd., the mining machinery manufacturer in Camborne, had, before the war, started an underground experimental near Troon in order to create a facility for proving and testing their

Plate 85. Surveying on the Main level by the turnout into the South Crosscut – C K Lim kneeling and Mr Head, lecturer, in white overalls - 1954

Plate 86.
Holman Test Mine surface drilling – 1950's

Plate 87. Drilling with a drifter, Test Mine – 1946

mines overseas. Mines working small narrow lodes, as in the last Cornish tin mines, were becoming rare then and hardly exist today. From shortly after the Second World War, Holmans staffed and ran practical drilling and blasting courses for CSM students. This was provided free, no doubt in the hope that generations of mining engineers would leave Camborne for the mining fields of the world with the name 'Holmans' firmly imprinted in their memories. In **Plate 86** one of the Holmans instructors watches over a student as he drills a hole with a 'sinker' into a granite block out in the quarry itself.

Plate 87, which appeared in the 1946 prospectus, shows a student drilling with a bar and arm drifter at the Test Mine. This was the standard way of drilling blastholes in a tunnel heading. The jack bar is set up vertically and supports a horizontal arm. The machine is mounted on a feed cradle which is attached to the arm by a machine clamp. As illustrated here the miner could drill a hole in any direction or inclination. He also could slide the machine and its clamp along the bar and, if need be, swing the arm. The level of bar can also be altered by sliding the arm up and down the jack-bar. The machine and arm were very heavy and there was a trick to moving the arm up and down. This was done by swinging the machine into the vertical position, pointing downwards and a piece of timber or steel placed between the machine and the ground. The arm could then be loosened and jacked up using the hand feed screw as a jack. Whilst the machine was lighter, technically things had not moved on much since the early years of the century – compare this photograph with Plate 29 that was taken some 40 years beforehand.

One of the advantages of using the Test Mine was that it gave the students the opportunity to see and

equipment. It was also used as a show place where they could demonstrate their products, in a working environment, to prospective customers. The mine was developed as a series of horizontal tunnels driven into the face of an abandoned granite quarry. Most of the tunnels were of a large cross-section, typically 9 ft. x 9 ft. or larger, which were more typical of the large

Plate 88. Drilling with a Handyrig,Test Mine – 1946

Plate 89. Drilling in South Crosscut – Hedley Clayton – circa 1968

use modern equipment. **Plate 88**, is a case in point. Here, in 1946, a machine has been mounted on a simple leg. This was the forerunner of the pneumatic air leg (or jack-leg as it sometimes referred to), which replaced the bar and arm system about 1950. A modern jack-leg machine can be seen in **Plate 89** taken about 1969 in the South Crosscut at Great Condurrow. The man on the left is Hedley Clayton[1]

Another attraction of the Test Mine **Figure 17** was the scale of equipment that could be used compared with Great Condurrow. **Plate 90** shows a scraper-loader in use at Holmans in 1948. This machine was used for cleaning development ends and was the forerunner of the rocker shovel which was introduced in the 1950's, and still in use today for mucking tracked haulages. This scraper-loader was a noisy, dangerous, inefficient machine but was still there well into the 1960's.

1 In 1975, when Tony Brooks was underground manager of the Baluba Mine in Zambia, Hedley Clayton was his assistant underground manager. Hedley Clayton died in a road accident in Zambia in November 2000.

On surface, in the immediate post-war period, little changed at King Edward, and the scene was much the same as it would have been in the 1930's.

The staff at the mine in March 1954 consisted of:

Name	Position		Weekly wage £	s.	d.
Arthur Roberts	(**Plate 91**)	Mine Foreman	6	10	0
Vivian Thomas	(**Plate 92**)	Engineer	6	15	0
"Sandy" Powell		Smith	5	17	6
Tom Mitchell	(**Plate 93**)	Miner	6	0	0
H Ford		Carpenter	5	17	6
Birt Yeo	(**Plate 94**)	Mill	6	10	0
V M Mitchell		Office Caretaker	1	17	6

Plate 91. Arthur Roberts, Mine Foreman – 1951

Plate 90. Scraper loader at the Test Mine – 1954

Plate 92. Vivian Thomas, Engineer – 1951

Tom Mitchell, **Plate 93** taken in 1951, is seen standing in front of the dry. Note that he is wearing the old fashioned 'felt' hat carrying a carbide lamp. Students at that time still continued to wear jackets and ties to lectures. Motor cars, for students a rarity before the Second World War, were far more common by this time.

One dramatic event, fortunately captured on film **Plate 95**, was when the 'new' winder house and the compressor house burnt down in 1957. The winder and the compressor had been stripped out years before. In the foreground are three water tanks then used to supply the mill. They were originally part of Willie Thomas's cyanide plant. This photograph was one of a series of the fire taken by a student (Courtenay Smale) who was denied permission by the School to have them published in the local paper.

In the mill, the stamps were only run infrequently for demonstration purposes. At this time Frank Michell was head of ore dressing. The Michell family of Redruth had been involved in Cornish mining and engineering since the 18th Century. Frank Michell had studied at the School of Mines

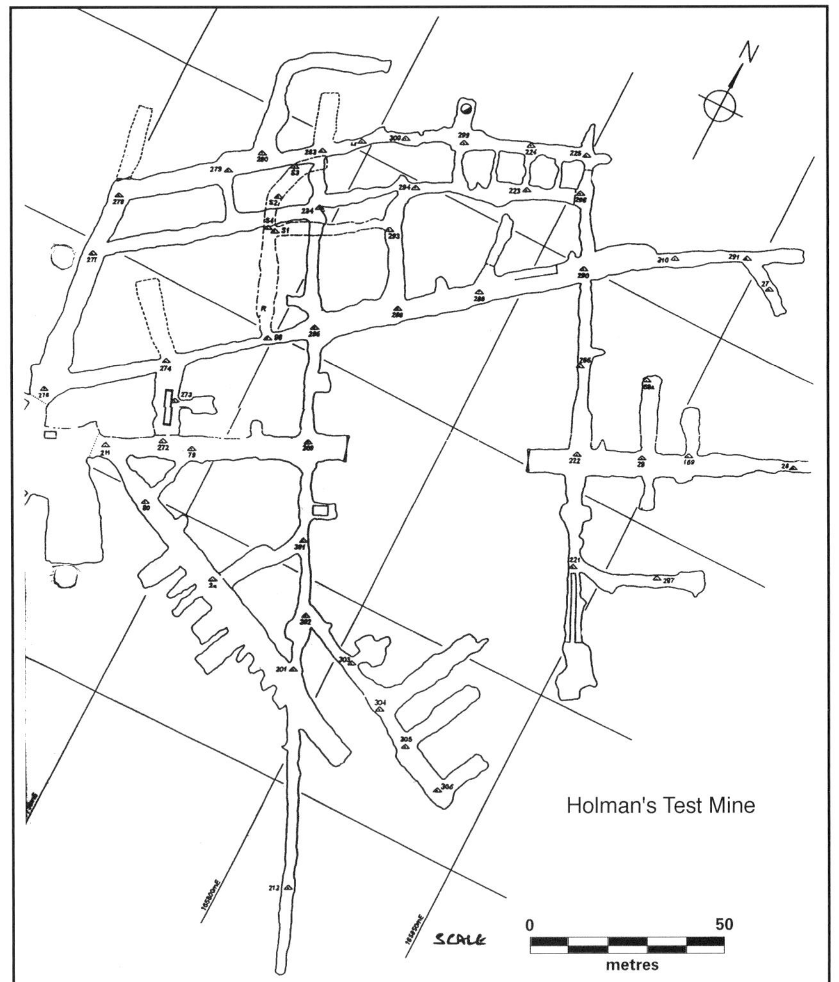

Figure 17. Plan of Holman Test Mine – 1998

Plate 93. Tom Mitchell, Miner – 1951

Plate 94. Birt Yeo, Mill man – 1951

and he followed in his father onto the staff of the school. He was not the most charismatic of lecturers, but he certainly understood mineral dressing, and some of his students went on to successful careers in this side of the industry.

Another character in the mill during the 1950's and 60's was "Jack" Keverne. **Plate 96**. Jack, whose father had worked at the Dolcoath mine, was born in Troon and was in the same year at the college as Michell. He worked extensively abroad before returning to CSM where he was the mill technician. He was known for his dry humour and his almost legendary prowess with a vanning shovel.

Plate 95. Winder house fire – 1957

Michell instigated many small scale changes and improvements in the mill. The mill had ceased to be a 'working tin plant' and could better be described as a experimental pilot plant with small size equipment that could be re-arranged to set up circuits capable of treating small samples of a wide variety of ores. The top end of the plant near to the stamps remained largely unaltered except that the Frue vanner was replaced by a third full size shaking table. Traditional methods continued to be taught and in **Plate 97** Arthur Roberts is introducing two overseas students to the mysteries of assaying using the vanning shovel. (Appendix II)

Externally the site remained largely unchanged. **Plate 98** shows the south end of the mill and the redundant calciner now used as a laboratory. Note the pipe chimney alongside the calciner stack. This came from a coal stove used to heat the building. King Edward can be a cold, wet and bleak place in the winter. Compare this photograph with Plate 52 taken from a similar position almost 60 years before.

George Whitworth retired in 1960 and he was replaced by a South African mining engineer, Ralph Gorges. During his reign cash was tight and there was little money for major additions at the mine, though a Craelius surface diamond drill was acquired for

Plate 96. Jack Keverne, Mill Technician – 1974

Plate 97. Vanning with Arthur Roberts – 1950's

demonstration purposes. During this period the survey office was lengthened by adding an extension to the western end, and the site of the South Condurrow dressing floors was levelled in an attempt to create a playing field. Sadly Ralph Gorges died in harness on the 16th of June 1969.

Plate 98. Lower mill section and calciner – 1957

Chapter Nine
The End of the 20th Century 1970-2000

Dr Peter Hackett, a mining/geotechnical engineer with an academic background, was appointed Principal in September 1970. The beginning of the 1970's heralded the start of changes that were to prove more significant than anything that had happened since 1921. The face of technical education generally, and the School of Mines in particular, was changing dramatically. For over 70 years, CSM had been a specialist mining college running one main programme of study – a 3 year diploma in mining engineering. Over the next few years, under pressure to expand and also to meet the changing demands of industry, CSM was to replace the mining diploma with an honours degree, and to bring in other degrees and diplomas in mineral processing, surveying and geology.

Since its inception, the School had its main teaching facility in the centre of Camborne (The School of Mines) with much of the mining, surveying and mineral dressing taught at King Edward. The Camborne facility was small and did not have the potential for the anticipated expansion that the School was looking for. A new, large, modern facility was needed. This was achieved when a new School of Mines complex was opened on the Trevenson campus at Pool, between Camborne and Redruth. The effect on King Edward was dramatic. Almost all of the mining and surveying teaching was transferred to Trevenson along with the mill pilot plant. At a stroke the whole mill complex at King Edward became redundant. On the other hand the expansion in courses and student numbers increased, at least in the short term, the demands for practical work in surface surveying and underground mining.

The mill, at the time of the move in 1974, was recorded photographically. The top end of the mill had not changed much in 20 years. **Plate 99** looking across the tables we can see

Plate 99. Holman-Mitchell flotation table – 1974

Plate 100. General view down the mill – 1974

that the full size sand table has been replaced by two half-size tables. In the centre is the Holman-Michell table. Michell had the idea of trying to separate the dense sulphides from the tin during tabling operations. The method was reasonably successful and some tables of this design were used for a period in commercial plants. The pulveriser is in the right foreground. **Plate 100** looking down the mill shows a mass of laboratory scale equipment and presents a very different picture from when it was an operational tin mill. The round frame which used to stand in the centre of the photograph is long gone, and the buddle has disappeared underneath a laboratory that had been created by partitioning off a section of the main mill area. Everything shown in this photograph was to be removed.

Aficionados of the Poldark television series, will remember that the early episodes had a strong mining theme. The mining scenes for one of the early programmes were shot underground at Great Condurrow. Despite the lack of Equity cards, likely looking members of staff and students were employed as extras. **Plate 101** was taken during location shooting in 1975.

The following plates, taken in 1976, by John Watton, were not for publicity purposes but rather to explore

Plate 101. Filming the Poldark Series-1975 – Tom Bowden, ?, Tony Batchelor, Dave Stark, John Watton

Plate 102. Tramming on the Main Level, Gt Condurrow – 1976

the techniques of underground photography. The quality of these photographs should be compared in style and content with the Burrow's photographs taken some 75 years before. All of these were taken on the main level at Great Condurrow using modern lighting methods and film.

Plate 102. Tramming an empty wagon back from the shaft. Looking west, East Shaft is to the right with the bottom of the escape ladder in the centre. The light coloured jack-bar on the right was used for surveying in the main level from wires suspended in the shaft. The South Crosscut is just off camera to the left of the picture. Between the wagon and the survey jack-bar is a hatch protecting the rock pass down to a small skip loading point in the shaft. The student on the left is John Eagling.

Plate 103 taken a few minutes later from almost the same point. The wagon is being turned into the South Crosscut. Note the very short wheelbase of the wagon that permits it to take sharp curves.

Plate 104 Pulling a chute. Compare this photograph with Plate 33 which is almost identical.

Plate 105 Surveying technology had moved on. Here a student is using a gyro-theodolite. In essence what this instrument does is to give a very accurate bearing. Getting a bearing, or direction, underground is difficult. The co-ordinates and elevation can readily be transmitted down a vertical shaft by means of a single wire. The problem is then to be able to orientate the workings accurately with the surface. This is of prime importance if one wishes to drive a tunnel from one shaft to another. Compare Plate 105 with Plate 85, the same chute is in the background

Mining is always changing. In the 1970's it was recognised that many of the Cornish mines had a

Plate 103. Tramming on the Main Level, Great Condurrow – 1976

Plate 104. Pulling a chute Great Condurrow – 1976

radon problem. Radon is the radioactive gaseous decay product of uranium. This was well known in the uranium mines around the world, and exposure to high levels of radon have been demonstrated to cause lung cancer. Most of the tin mines in Cornwall worked lodes that occurred in granite which contains traces of uranium and this is the source of the radon. The gas can be controlled by good ventilation. Here, in 1977, **Plate 106** Tony Brooks is carrying out a radon survey using an Instant Level Working Meter.

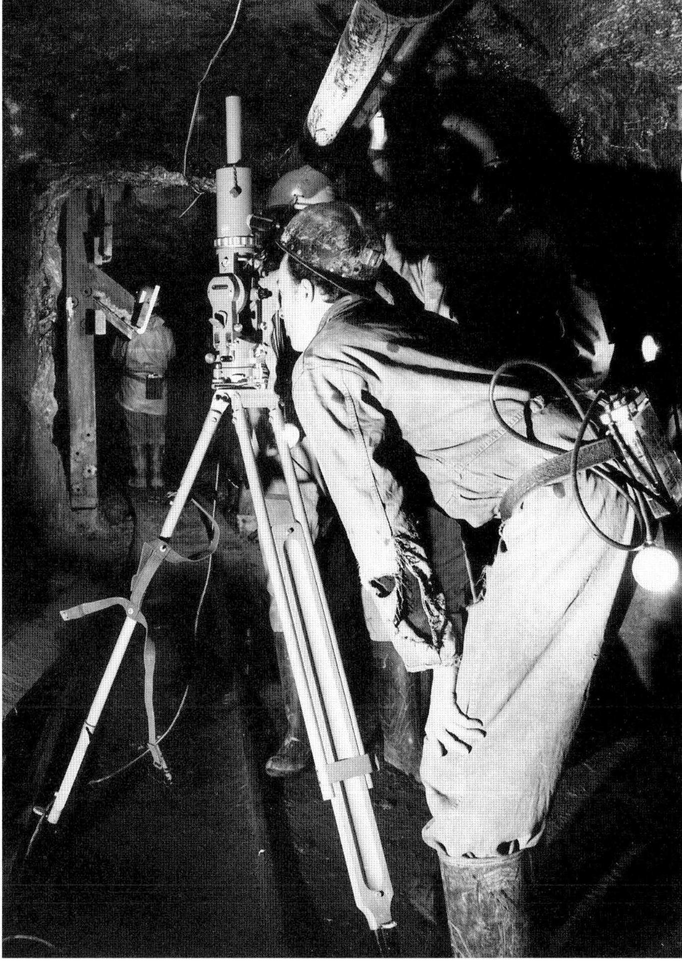

Plate 105. Underground surveying with a gyro-theodolite – 1977

Plate 106. Radon survey, Tony Brooks – 1977

A measured volume of air is drawn by a pump through a special filter paper which traps the radio active products (minute solids) of the decay of radon. The filter is shown here being put into the meter for counting. From the measurement it is possible to calculate the concentration of radon in the mine atmosphere.

In the old western portion of the mine the Mine Captain, John Bullock, and his assistant, the late Arthur Dodd, are making their way up the ladders towards West Shaft and surface after a hard afternoon underground. **Plate 107**. The lode had been completely stoped away here. Note how the timbering is supporting the ladderway and the platforms.

Plate 107. Climbing to Grass, Arthur Dodd and John Bullock – 1983

Chapter Ten
Restoration of the Mill

By the mid-1980's part of the site which included the mill, stamps, mill engine house, boiler house and the calciner had been largely stripped of their equipment and they were no longer required by the School of Mines. The historic importance of the site was not then fully appreciated and there were suggestions that some of these structures be demolished. At that time conservation of mining buildings and sites in Cornwall was very much in its infancy.

There were no mining museums in the Camborne-Redruth-St Day mining area – an area that was, in the middle of the 19th Century, one of the most important mining areas in the world. In 1987 the School were looking at ways of broadening its marketing image and one suggestion, made by John Watton, was to turn the redundant mill complex at King Edward into a museum. A volunteer group was set up to try to achieve this. The project was supported by the School of Mines and, initially, most of the volunteers came from the School's staff.

The objectives of the group can be summarised as follows:

i) To preserve the buildings and the site which is of significant historical importance.
ii) To re-equip the mill to working condition using, where possible, rescued and preserved equipment which itself is of historical interest.
iii) To establish a small museum telling the story of King Edward Mine, the local "Flat Lode Mining Area", mining methods and tin dressing..
iv) To rescue and to preserve industrial plant and equipment relevant to Cornish mining industry.
v) To develop the site as a visitor centre serving the west end of the Mineral Tramways network, which has a route passing through the site.

Work started first in the mill engine room which was to be developed as the museum.

A number of redundant show cases were acquired from Torquay Museum and from the Royal Geological Museum in Penzance. After some refurbishment these cases were erected in the museum and some rudimentary displays constructed. The volunteer group's interest soon moved to the mill which offered real challenges of acquisition, engineering and improvisation.

With the restoration of the mill there were two possible strategies. The first would have been to have attempted to replicate the machinery and flowsheet exactly as it was during the first decade of the 20th Century. This would have been very expensive and would have involved a significant amount of construction of complex machinery from scratch where originals no longer existed. The other route was to try to replicate the mill as far as was practicable using machinery that could be rescued or salvaged from existing and derelict mine sites in Cornwall.

The second route was the only one open to the volunteer group which. at that time, had no money and no right of access or tenure to the site. With hindsight this was the best way forward anyway. The almost complete closure of Cornwall's tin mining industry, provided the opportunity to rescue and preserve machinery that would otherwise have been lost. The mill as now constructed is equipped with real equipment which has a real history. The flow sheet as re-constructed is almost identical to that put together by William Thomas. The Buss table has been replaced by a Holman sand table and the Frue vanner by a Holman slime table. As far as we have been able to establish no Buss table or Frue vanner has survived anywhere in the world, although there is a suggestion that a Frue vanner has been salvaged in Tasmania. However, nothing that has been done is not reversible. If, in the future, a Buss table or a Frue vanner turn up then they could be installed instead of the present tables.

Soon after restoration work started, the volunteer group were encouraged by two events. The first was in 1989 when the Cornwall Archaeological Unit produced its splendid report titled the Central Mineral Tramways. This report outlined the concept of connecting the mining/industrial sites of central-west Cornwall through footpaths utilising the now disused mineral tramways and railways. King Edward Mine was considered to be an integral part of this concept. At much the same time the site was visited by an inspector from English Heritage. The outcome of this visit was that all of the buildings on the site were Listed Grade II*. This very high listing indicates that they are of national importance and are in the top 5% of listed buildings in the country.

Plate 108. Interior of the mill – 1987

When restoration started the mill had been almost totally stripped and was being used as a store. **Plate 108** was taken looking down through the mill in 1987. In the middle distance is an Atlas Copco LM36 rocker-shovel being rebuilt prior to transfer to the former Holman Test Mine. This should be compared with Plate 100 taken from a similar point 13 years earlier. Fortunately the magnificent set of Californian stamps were still in position as was some of the water pipe work. **Plate 109** also taken in 1987 looking back up the mill towards the stamps There was no electric power or lighting.

Whilst there was an overall general plan of how we would have liked to have seen the mill, in reality the re-construction was driven by what machinery could be found. Back in 1974, the Trevithick Society had rescued two Cornish round frames when the stream works at Tolgarrick in Tuckingmill was being cleared. These had been stored in the old Vounder china clay dry near St Austell and were made available to King Edward. **Plate 110** illustrates what two round frames look like in 'kit form'. The parts, which consisted of two small lorry loads of pieces of wood, had been carefully numbered when dismantled. Unfortunately the key to the numbering had been

Plate 109. Stamps – 1987

Plate 110. Dismantled round frames in Vounder clay dry – 1988

lost! One frame was made out of the pile of pieces and erected exactly where the original round frame had stood – utilising the old centre support which still remained in the mill floor. **Plate 111**.

The foundations of two of the full size shaking tables still existed. It was decided to use these for a sand table and a slime table. A complete sand table, originally acquired from the old South Crofty mill by Marine Mining, was donated by Mr Mike Proudfoot (a former student). **Plate 112** shows this sand table in process of being erected late in 1989. Later a slime table, recovered from Tolgarrick tin, with a deck from the Geevor Tin Mine, was put up beyond the sand table.

The old foundations for the half-size table were no longer required and had to be removed. **Plate 113**

Plate 111. Round frame nearing completion – 1991

Plate 112. Erecting sand table – 1989

shows Frank Kneebone attacking the concrete of one of the foundations with a Holman paving breaker. Further down towards the bottom of the mill another modern concrete foundation had to be removed. This resisted all attempts to break it up using the breaker and as a final resort high explosives had to be used. The foundation was drilled, lightly charged with gelignite and blasted. Remarkably, this was achieved without breaking any of the windows or blowing the roof off the building, although some of the protective sandbags were found later wedged up in the rafters.

The table foundations were removed to make way for a 14-ft diameter double dipper wheel recovered from the Tolgus Tin site where it was about to be

Plate 113. Frank Kneebone removing half table foundation – 1989

Plate 114. Building the 14-ft dipper wheel, John Goodman and Frank Kneebone – 1992

Plate 115. Erecting the Hardinge mill – 1993

scrapped. **Plate 114** has John Goodman and Frank Kneebone bolting on one of the buckets. This was set up such that one side of this wheel would elevate the table middlings to the ball mill and the other side to return the product from the ball mill back to the stamps classifier.

An old Hardinge ball mill was acquired from the outfit then scrapping the mill at Geevor. **Plate 115** and it was erected on new foundations erected where the old pulveriser once stood. There was no lifting plant in the mill and all heavy machinery had to be moved by the traditional methods of muscle power, levers and jacks. The mill, weighing well over a ton, was literally rolled into the mill, the settling pits and launders being bridged on the route across the mill. Similarly, an old batch flotation cell, dating from about 1930, was recovered from Geevor. Here **Plate 116** shows it being transported into the mill on an old mine materials wagon. In tin dressing flotation replaced the inefficient and environ-mentally damaging calcination for the removal of sulphides from the concentrate[1]. These old cells are very rare and this one was repaired and installed in a space that had been occupied by a second pair of kieves.

It was realised that the 'modern' laboratory had been built over the site of the convex buddle. The structure was demolished **Plate 117** seen here in a state of controlled destruction. The buddle was found to be largely intact having been filled in below the modern concrete floor, just about where the ladders are in the photograph. The stripping out of the laboratory also exposed the area where the kieve tossing and packing gear used to stand. A partially complete set of gear, possibly the only one in the world, was found at Geevor in 1989. **Plate 118** shows this gear mechanism in the process of re-erection. This can be compared with the original installation Plate 56

By 1999 much of the installation work in the mill had been completed and the stamps were run and tested. The roar of the stamps was heard at the mine for the first time in over 45 years.

Engine Shaft, which had collapsed and had been filled in many years before, was considered to be stable and safe. This assumption was proven to be spectacularly incorrect when, on the 19th January 1994, the shaft opened up to surface. **Plate 119** In the background is the mill engine room and, in

Plate 116. Bringing the batch flotation cell into the mill – 1992

Plate 117. Demolishing the mill laboratory – 1989

1 Successful flotation depends on adding to the pulp a mix of chemicals designed to make selected minerals hydrophobic. Sulphide minerals lend themselves to this. Air is then bubbled through the pulp. The coated sulphide minerals adhere to the resulting bubbles and float to the top where they can be scraped off. Conversely the cassiterite, which is very difficult to float, remains in the cell and is drawn off from the bottom.

Plate 118. Installing the tossing and packing gear – 1991

Plate 119. Collapse of Engine Shaft, 19th January 1994

the distance, the ruined South Condurrow stamps engine house. Over the following weeks the hole expanded until it was some 25 ft. wide and 25 ft. deep and was threatening to undermine the mill engine room.

This shaft, and three others on the site, were stabilised by Kerrier District Council under their shaft capping programme. All were capped in a way that left an opening into the shafts. At Engine Shaft, this was done by setting a concrete slab 30 ft. down on stable rock leaving an access through the slab into the choked shaft below. **Plate 120** shows the excavation down to rock head. On the far side of the excavation can be seen the remnants of the bob-wall of the 19th Century pumping engine house. Concrete rings were extended above the slab to within 12 ft. of surface. The last 12 ft. was completed as a full-size rectangular shaft. It is planned to erect a headframe over the shaft and install three or four shaft sets in the newly created collar.

In March 2000 the South Condurrow Stamps

engine house, which had been abandoned since 1914, was surrounded by scaffolding during stabilisation. It was from here that **Plate 121** was taken in March 2000. Comparing this with **Plate 122** taken some 90 years before hand – it is remarkable how little has changed

Up to the end of 2000, most of the work of restoration had been carried out by the volunteer group with a limited amount of material assistance from the School. As part of the overall Mineral Tramways strategy, European Funding was granted to assist in bringing King Edward up to a standard where it could be opened to the public. This was a major step forward and something that the volunteer group had always hoped for.

Contractors started on site in February 2001. The contract included much repair work on the fabric of the buildings including total re-wiring, refurbishment of the toilet block and the creation of an entrance, reception and retail area in the calciner and old laboratory at the south end of the mill.

Along with the building works the opportunity was taken to uprate the museum, though it might be better described as an interpretation centre.

The display cabinets were refurbished and a new set of displays developed from scratch. These displays are not just a collection of articles, but an attempt to demonstrate how Cornish mining and tin dressing worked in the Edwardian period. This has been

Plate 120. Excavation of Engine Shaft during stabilisation – 1996

Plate 121. Surface view of King Edward – May 2000

Plate 122. Surface view of King Edward – 1906

done taking King Edward as the model and utilising many of the photographs used in this book.

The King Edward Project has been all about people, and in particular the many volunteers who have worked there on every Sunday over the years. Some have come for a month, some for a year and some for over a decade. King Edward as it stands today is a reflection of their efforts, application, enthusiasm and, above all, their humour.

Plate 123, Peter Forbes, caught in early morning sunshine – when asked why the box that he was making resembled a coffin, replied: *'This job will be the death of me'*

Plate 124 Willie Uren and Peter Forbes making a collecting box

Peter *I think that is near enough.*
Willie *Got to be exact*
Peter *'Tis exact*
Willie *Well, 'tis near enough then*

Plate 123. Peter Forbes "This will be the death of me" – 1996

Plate 124. Peter Forbes & Willie Uren "Got to be exact" – 1996

Plate 125. Aerial phototograph looking north – about 1993

Chapter Eleven
Photography at King Edward Mine

Early Underground Photography

In considering the images captured at King Edward Mine an understanding of the historical background leading up to the first underground illustrations of the mine will place the work in context.

The taking of pictures underground dates from 1861 when Felix Nadar succeeded in taking photographs using arc lamps in the catacombs in Paris.[1] The arc lamps were powered by 50 Bunsen cells, Nadar having first used this set up in his studio about three years earlier.[2]

By the time the first underground images were made in Cornwall (1865) Professor Robert Wilhelm Bunsen (of Bunsen burner fame) in collaboration with Sir Henry Roscoe had demonstrated a new illuminating agent. This agent was magnesium.[3] It was loaded as a thin ribbon into a clockwork apparatus that gradually transported it into a spirit flame where it burned with a brilliant light.

The first underground pictures taken in Cornwall were by W.E. Debenham whose studios were in London. Debenham's images were taken at Botallack Mine near St.Just, where Robert Preston had photographed the Royal visit of the Prince & Princess of Wales in July of that year. Debenham used wet plates to capture the images with magnesium ribbon providing the light. The wet plates he used had been invented by Frederick Scott-Archer some 14 years earlier in 1851. These plates superseded the Daguerreotype & Calotype processes. The wet plates had a number of advantages over these earlier systems – namely fine grain, good contrast, quick processing and they could be dried by heat. However wet plates had a number of disadvantages especially for those photographers working in caves or mines. The plate had to be immersed in silver nitrate after being coated with iodized collodion immediately before exposure. This together with other chemical solutions to process and fix the plates had to be carried underground and was a considerable inconvenience.

1865 was a busy year for photographers making underground images, with Alfred Brothers working in the Blue John Caverns in Derbyshire, Henry Jackson recording scenes in the Bradford Colliery, and Debenham at Botallack. Photography underground was thus established in various areas of the United Kingdom almost simultaneously. However the very slow speed of the photographic plates and the fact that magnesium ribbon burned very slowly meant that the exposures were lengthy. This also meant that men could not be recorded at work since any slight movement would spoil the plate.

At this time (1865) due to these difficulties it was inevitable that underground photographs were nothing more than record photographs. Where

1 Newall B, The History of Photography. New York. 1982.
2 Howes C., To Photograph Darkness. Gloucester. 1989.
3 Watton W J., The History of Underground Photography in Cornwall. The Trevithick Society Journal, 1990.

required, mannequins would be used to depict the men at work, some Nadar images prove this to be the case. What was needed was a means of producing a lot of light quickly. Experiments using powdered magnesium and various oxidizing agents (to make it burn quickly) were performed over the next 20 years. Accidents were common with mixtures such as magnesium and gunpowder!

Johannes Gaedicke and Adolph Miethe solved the problem in 1886, when their experiments using powdered magnesium and potassium chlorate were highly successful. Using flash powder in large quantities was still dangerous and there were five fatalities during the period 1887-1890. Flash powder[4] was generally used in a metal tray, the powder being ignited by a spark from a wheel and flint or by touch-paper previously lit by a match. Usually these touch papers were filter papers that had been previously soaked in potassium nitrate and then dried. The burning flash powder mixture gave off a great deal of smoke which could rapidly fill an underground chamber making any attempt at a second exposure useless until the combustion products had been cleared by whatever ventilation existed.

The change from wet plates to dry plates took place in the mid 1880's and with it underground photography became somewhat easier. The new dry plates were still extremely slow in speed by modern standards, about 4 ISO being the norm at that time.

The Photographers

John Charles Burrow, Herbert Hughes and William Thomas (Plate 1). These three are considered together for, as we shall see, they often worked together in Cornwall.

During the spring or early summer of 1891 William Thomas suggested to his friend and local photographer John Charles Burrow[5] that he might like:

the idea of extending his experience in mining photography to underground workings[6]

William Thomas's prime interest in the photography was the acquisition of a set of lantern plates which could be used to illustrate his lectures. William Thomas was competent photographer in his own right. A photograph album of his images has survived and can be seen at the Cornish Studies Library in Redruth. This album also contains work by N.Ward and Burrow's son, H.A.Burrow.

Prior to this invitation Burrow wrote:

I have often been entrusted with the preparation of photographs of mines and mining appliances (on surface) for lantern or other educational uses at the Camborne Mining School, for book or catalogue illustration and for general purposes

William Thomas suggested to Burrow that he should make contact with other photographic image makers who had attempted underground photography with a view to finding out how they achieved their results. To this end Burrow wrote to Arthur Sopwith a colliery manager from Cannock Chase together with another colliery manager Herbert Hughes[7] of the Lye Cross Colliery in Staffordshire. These photographers had already been recognised for their efforts in this field of photography by being awarded medals from various societies. Burrow's research also included writing to Heinz Börner who was known for his photographs taken in the Freiberg mines. Thomas had probably seen Börner's work when he visited Germany in 1891. Börner sent copies of his work together with

4 In 1896 the Ilford Manual gave "A good mixture for flash powder" as follows:- potassium perchlorate 3 parts, potassium chlorate 3 parts, magnesium powder 4 parts. The perchlorate and chlorate may be finely powdered and mixed together with the magnesium added only when ready for immediate use. A note of caution was added in the manual stating that this was extremely dangerous when used in large quantities and likely to cause explosions.

5 J.C. Burrow was born on 11 March 1852. He had started his photographic career in Truro with his brother Henry with a studio in Ferris Town. In or about 1873 he moved to Trelowarren Street in Camborne to start business on his own account. It was from this base that he quickly became recognised as a good general photographer gaining many clients in the mining and the associated industries.

6 Burrow J.C. and Thomas W., 'Mongst Mines and Miners. Camborne, Cornwall. 1893. Reprinted 1965.

7 Herbert W Hughes was born in 1863. He was the son of a mining engineer John Hughes.

detailed information on the lighting techniques that he had used underground.

Burrow compared the work of these three photographers. It was obvious that the images made by Herbert Hughes were far superior to those of Arthur Sopwith. Craft skills used by Hughes were different and more creative. The lighting techniques employed by Herbert Hughes must have been exceedingly dangerous in some coal mine situations where flammable gas may have been present. His lighting involved the use of powdered magnesium for the main illumination. This was burnt in an oxyhydrogen flame with a reflector to concentrate the rays of light onto the subject. This apparatus was marketed by the Platinotype Company and was essentially an updated version of the old limelight burners developed by Thomas Drummond in 1826 which were based on the invention of limelight by Goldsworthy Gurney in the early 1820's[8].

The correspondence between the two photographers (Burrow and Hughes) resulted in Hughes making a visit to Cornwall to work with Burrow and Thomas. This was destined to become an annual event and Burrow remarked during a lecture to the Scientific Societies of Cornwall in 1894 that; *Hughes has become a member of my gang.*

Both men adopted similar technical cameras, Burrow choosing half plate format – whose images measured 4.75 ins. by 6.50 ins.. Burrow's preference for make was the Thornton Pickard 'Ruby' model. This camera would fold down into a quite compact unit considering the size of negative produced, with obvious advantages for transportation into difficult places underground.

Hughes choose a half plate camera made by Underwood & Co. and a whole plate camera made by Watson – whole plate format is 8.5-ins x 6.5-ins. He preferred the Watson Acme whole plate camera for underground photography. This was because most of the movements are regulated and locked by spring catches and the preliminary operations are conducted in semi-darkness – it is an advantage to know from the sound given by the catches when the several parts are pushed into a fixed position[9]

There are 14 whole plate negatives of King Edward Mine at the Royal Institution of Cornwall which almost certainly, due to their date, were taken by Herbert Hughes.

Burrow commenced his experiments in underground photography in the autumn of 1891 at South Condurrow Mine. As we have seen, students from the School of Mines did some of their practical training there. At this time William Rich was the mine's Chief Agent. Rich was a friend of both Thomas and Burrow. Thus it was perhaps easier for Burrow to gain access to this mine rather than some of the others in the area. His early attempts at this work were encouraging although not entirely successful, with only about 15% of his negatives being printable. He used Tyler's double dark slides carrying a total of 14 dry plates to the scene of operations. Due to the difficulties encountered in working underground he usually only exposed 5 or 6 plates in a whole day's work.

Burrow recorded much of this experimental information and presented it in a lecture to the Scientific Societies of Cornwall held in Penzance on the 15th of June, 1894. It was published by the Royal Geological Society of Cornwall later that year. Burrow's presentation was illustrated by lantern slides made from photographs taken in German, Cornish, and other English mines.

The single most important item was obviously the lens. Burrow informs us that the very best was found to be a Zeiss Anastigmat Series III, manufactured by Ross & Co.. He also records details of his experiments with different photographic plates, finally selecting the Cadett Lightening Plate. This plate gave him the least

8 Porter D.H., "The Life and Times of Sir Goldsworthy Gurney", Associated University Press 1998.
9 Hughes H.W., Photography in Coal Mines, The Photographic Journal, London. 1893.

trouble with the men's candles when recording images with them lit. The speed of the lighting plate was H.D.250 (Hurter & Driffield speed system) which is equiv alent to about 10 ISO in the speed rating we use today.

The experiments undertaken at South Condurrow Mine with regard to lighting convinced Burrow of the need to create as much light as possible underground. This would be particularly important since he intended to broadened his horizons and commence image making in the other larger mining concerns nearby. Reports by the agents and managers of some of these mines talked of champion lodes 30 feet across and of huge gunnises where the lode had been removed. Such reports convinced him that the normal flash powder trays would be totally inadequate. He therefore set about designing a pair of 'Triple' magnesium flash trays, to be manufactured by William Rich's engineering factory in Redruth. **Figure 18** Harry Rich, William Rich's son, was also a very accomplished amateur photographer. His connections through mining engineering with his father's business eventually led to him becoming a Governor of the School of Mines.[10]

WILLIAM RICH & SONS, REDRUTH,

MINING ENGINEERS AND MINING MERCHANTS.

(Established 1876). Telegrams: "RICH, REDRUTH" (A B C Code used). 'Phone—REDRUTH 65.

AGENTS—Nobel's Explosives and Various Manufacturing Co.'s.

MACHINERY AND MATERIALS ON SALE:—

Rockbreakers,	Hoists, Kibbles, Skips,	Ore Bags, Barrows,	Carbide, Lamps, Candles,
Grizzleys,	Tram Wagons, Rails,	Underground Hats,	Tools (Shovels, Sledges,
Pumps, Boilers,	Ventilating Fans,	Boots, Overalls,	Picks, Dubbers, etc.)
Stamp Batteries,	Tubes and Fittings,	Explosives, Fuses,	Handles for above (Ash,
Shoes, Dies, Grates,	Galvanized Sheets,	Blasting Accessories,	Hickory),
Calciners,	Wire and Hemp Ropes,	Belts:—(Vanner,	Engineers' Sundries,
Concentrating Machines,	Lifting Tackle, Chains,	Driving, etc.)	General Quarrying and
Tube Mills, Flints,	Anvils, Forges,	Oil, Grease,	Mining Requisites,
Steam and Gas Engines,	Steel, Iron, Nails,	Heating Apparatus,	Second-hand Appliances.

Figure 18. William Rich advertisement – 1920

Rich's engineering factory duly produced the triple magnesium flash-trays to Burrow's design and they were complete with parabolic reflectors to direct the rays of light towards the subject. These lamps often produced dense areas of shadow, so to lighten this to enable detail to be seen (or 'fill' in the shadows as it is known) Burrow used limelight burners. These were powered by Brin's cylinders of compressed gas. In **Plate 126**, taken on the 400-fm level at Dolcoath, the cylinders and burners of just such a set can be seen on the right. Thus the preparations to descend into South Condurrow – or for that matter into any mine – were extensive.

Burrow had a 'gang' of helpers for these photographic excursions underground. One person adjusted the limelight burners as well as transporting them to each location. Another two carried the oxygen and hydrogen cylinders needed for the limelight burners. There were two further assistants involved with the transportation of the triple magnesium burning trays, the reflectors, and stands. They were also responsible for the magnesium and oxidizing agents which they mixed on site underground for

10 Harry Rich's lantern projector and some slides were offered for sale some years ago, and now are part of John Watton's collection.

Plate 126. 400-fm level Dolcoath, showing gas cylinders for limelight burner – about 1895

safety reasons. One last helper was responsible for the tripod, plate carriers and other photographic paraphernalia. Burrow transported the camera, dark-cloth and lens.

South Condurrow was an ideal testing ground, where the temperature remained fairly constant. Some other mines in the area nearby which were mining at much greater depths gave the photographers real problems. Mines such as Dolcoath and Cooks Kitchen were in this category where the camera equipment would be rendered useless by a film of condensation for quite a long period of time after the location for a shoot had been reached. Burrow attempted to counteract this problem by carrying the lens tucked inside his shirt next to his body. Hughes adopted a similar principle by placing the lens in a trouser pocket to keep it at or near body temperature.

Working with Thomas in difficult conditions Burrow quickly learnt various aspects of hard rock mining and he described some of the difficulties encountered:

Some of the chief difficulties that exist to the photographer are dripping or running water, vapour, mud, grease, damp smoky air, heat, cold, draughts, and darkness. The kid gloved photographer with his delicate apparatus, who is afraid to soil his hands or his camera had better not attempt photography in our deep Cornish mines. The photographer should in fact be something of a mining engineer, to intelligently grasp the idea which the picture is intended to illustrate.

The partnership between Burrow and Thomas extended beyond the activities of South Condurrow. In 1893 they published a book called "Mongst Mines & Miners". Burrow making all the illustrations and Thomas providing the text to compliment them. Both Burrow and Hughes joined the Royal Photographic Society in 1893. Both were awarded Fellowships in 1895, Burrow for his images taken underground in Cornwall and Hughes for submitting a series of coal mining images taken underground in the Midlands, and written work describing his experiences in this field of photography.

The photographs for the book by Burrow and Thomas were produced to an extremely high standard by the Woodburytype method. This is the process of printing with pigmented gelatin from lead moulds which, in turn, had been obtained from chromated gelatin reliefs. This is an Intaglio process where the density of the image controlled the depth of impression in the lead block, which, in turn held more or less ink thus determining the finished density of the print. This process was selected by Burrow since it produced prints with all the half tones, (the images were all reproduced as sepia prints), gave high definition, but required no dot screen.

Some of the later underground photographs produced by Burrow and Hughes are real pieces of art. Not only is the quality quite remarkable, quality that many photographers today would struggle to equal, but they also show a high level of artistic creativity. Plate 30, sinking a winze below the 30 fathom level, is a case in point. Here the main subject in the foreground on the left hand side are the three miners operating a windlass. This is lit, not from by the camera, but from a concealed point on the right. The eye is then drawn to the tunnel leading away to the right where a chute is illuminated by a second light out of sight round the corner. This secondary lighting creates a sense of depth. Clearly this photograph was 'designed'. Remember, he would only have had a few candles for illumination and he could not have seen, but only visualised, how it would have looked in the photograph. This should be compared with Plate 32, of a man tramming, taken by N.Ward at about the same period. This exhibits little creativity and is front lit by a single flash.

Burrow was appointed Royal Photographer to King Edward the Seventh and took up his post in 1901, when it appears that he moved to London. Previously he had a Royal appointment to Edward when he was Prince of Wales and was allowed to use the Prince of Wales insignia (feathers) on his stationery. This position did not necessitate his residing in London.

Burrow, Thomas and Hughes remained friends and when Thomas left the Mining School to become Manager of Botallack Mine in 1906 they made a series of surface illustrations at Botallack. These negatives may be found in the Cornish Studies Library in Redruth.

Burrow made excursions home to Cornwall for holidays and he spent much time with Hughes and Thomas on these occasions. They enjoyed taking photographs of various beauty spots in the County and the Isles of Scilly. This was now only surface work as, from about 1905, Burrow had been forced to stop underground photography, on the advice of his doctor, following heart problems.

This remarkable man, regarded along with Herbert Hughes as one of the fathers of underground photography in mines, died in London on the 25th of October 1914, aged 62. The West Briton, of the 28th October 1914, said that he was:

well known throughout the mining world as a photographer. He was manager of the Camborne Printing Co., Director of the Town Hall Building Co., and a prominent Freemason.

Following his death, Burrow's commercial property in Cross Street Camborne was rented by his nephew, Nankivell, a grocer. Nankivell had no use for the huge quantity of glass negatives which were stored at the rear of the building. The vast majority of these plates would, of course, be work of a general nature including weddings and portraits etc. However there must also have been a reasonable number of mining surface scenes, and the underground scenes for which he was famous. Shortly after taking over the premises Nankivell is reputed to have offered the mining plates to the School of Mines. The School apparently turned down his offer, though this is difficult to comprehend since they had been heavily involved with Burrow's work in this field.

There are many stories connected with the fate of these mining negatives and some people say they had their emulsion layers scraped off and were used to glaze a conservatory or a greenhouse. Indeed it is quite possible that hundreds of general negatives suffered this fate. Fortunately the important mining plates taken underground were removed by Herbert Hughes and taken back to his home in Dudley. Here he carefully stored them along with his own plates of mining scenes, until his death, in August 1937. The affairs of his estate were protracted and the 1939-45 war intervened before his family was able to return the plates to Cornwall. In 1952 they made a most generous gift of all of Burrow's negatives together with all of Hughes's Cornish underground negatives to the Royal Institution of Cornwall. Of these 39 are scenes taken in King Edward Mine.

Burrow's son, **Henry Arthur Burrow**, attended the Camborne School of Mines as did William John Bennett's son James. W.J.Bennetts was another well respected photographer who had connections with the mining industry. The photographers' sons became friends, and both attained First Class passes at the School. Henry Arthur Burrow tried his hand at underground photography at King Edward Mine, but he did not compare with his father.

William John Bennetts started a photography business in Hayle, but by 1895 was established in Camborne. Photography was only part of his empire since he also had a grocery business run by his wife Mary. They had eight children three girls and five boys. Two of the boys and two of the girls were later to be trained in the art of photography by their father, who had been trained in Plymouth prior to 1880. James Bennetts was one of the sons who was trained by his father but left the family business when he joined the Camborne School of Mines as a student in 1900.

Bennetts used Houghton Ensign camera equipment, consisting of cameras varying in size from 12 ins. by 10 ins. to half plate. He undertook mining photography contracts with South Crofty and Dolcoath, as well as extensive catalogue illustration for Holman Bros. Many Holman catalogues display his work as photographs, retouched photographs or drawings made from his photographic images. His work was somewhat unimaginative in its approach especially with regard to the lighting techniques employed. Thumbing through these same Holman catalogues today it interesting to note that nearly all the underground images are from the camera of J.C. Burrow. Bennetts or possibly his son, also took some underground photographs for the School at Great Condurrow. Plate 78, which appeared in the 1936 prospectus, showing students surveying is certainly by them and it is probable that Plates 76 and 77, which appeared in the same publication, are as well. Bennetts was responsible for a number of surface views of King Edward Mine. Bennetts work has survived in great quantities in the form of postcards one of which is shown in Plate 51 and as lantern plates where his collection, together with his lantern plate projector, is carefully preserved by Arthur Osborne of Camborne and a collection of over 800 glass negatives in the Cornish Studies Library, Redruth. Bennetts remained active in photography until a few months before his death in 1943 aged 94 years.

Thomas Roskrow who had a stationery business at 28 Fore Street, Redruth published a postcard in 1904/5, **Plate 127**, taken on the Great Flat Lode in King Edward Mine. It is by no means clear whether Roskrow was the author or if he engaged the services of another photographer to capture the image and then marketed it through his business. The image construction and 'record' lighting are suspiciously like the underground images of W.J. Bennetts, although it is highly unlikely that Bennetts would have allowed the card to be marketed without his name being credited. There were a number of other photographers with mining connections such as Chenhall from Redruth as well as several others in the Camborne/Redruth district who may have taken the image or shown Roskrow how to attempt it.

N. Ward is credited with Plate 32, but nothing is known about him. He was not a member of staff at the School, nor was he a student. The name is not listed in Kelly's Directories for the time 1900-1910. If any reader were able to supply any information about this photographer it would be very much appreciated.

There have been many attempts by various amateur and professional photographers to make images at King Edward. Little is known of the photographers who took the images included in this book between 1914 and 1950.

Gordon Nicholas joined the staff of the School of Mines in 1950 teaching mathematics and engineering drawing. He was a very good amateur photographer

Plate 127. Great Flat Lode by Roskrow – 1904

and was keen to accept the challenge of making images underground. He used a twin lens Rollieflex camera, with a Zeiss Tessar lens. Nicholas mainly used flash bulbs underground achieving differing intensities of light by the size of bulbs employed. Main illumination was usually from PF5 bulbs with PFI bulbs being used as fill in. He sometimes used a combination of bulb and electronic flash but did not find the latter reliable in areas underground where the heat and humidity caused condensation problems. These problems affected his camera equipment also, in particular the lenses of his Rollieflex were very vulnerable. Nicholas used a novel remedy to solve this problem, he made up a pad from a new lens cloth and fixed this in place over the taking lens until the very moment of exposure. The viewing lens of the twin lens camera could be wiped clear as and when necessary.

The procedure outlined above enabled him to take recruitment pictures for South Crofty mine in 1956. He was also responsible for another set of pictures taken for a similar purpose in Geevor mine in 1957, although this mine was cooler and presented little or no problems with condensation. It is quite possible he is the author of the plates taken for the School's 1954 prospectus (Plates 85 and 90). The School, always mindful of costs, sought to utilise the talents of the staff in the production of the prospectus and Gordon Nicholas would have been the obvious choice.

Bryan Earl attended Camborne School of Mines from 1950-53, being encouraged to enrol by a beautifully illustrated prospectus of some fifty odd pages bound in a glorious blue cover[11]. His connections with mining began in 1944 when, as a 'Bevin' boy, he was conscripted to work underground in various collieries in Kent, Derbyshire, and Leicestershire. Earlier, he had been permitted to use his father's darkroom to process monochrome films and this introduced him to photography at an early age.

Earl's grandmother had given him a Kodak Autographic folding camera, made in about 1920. Plates 91-94 were taken with this camera in 1952 on 116 size Verichrome orthochromatic roll film, the only film available to fit it in the 1950's he recalled recently.

Later, in 1952, he invested in a new Ilford Advocate 35mm roll film camera, this camera cost £25.00 which was the equivalent of about six weeks wages at that time. The Advocate had a 35mm f4.5 Dallmeyer lens which was wide enough for most applications underground. With this camera he was one of the first, if not the first, to attempt colour photography underground in Cornwall.[12] To attempt this colour work Earl used tungsten based film with illumination from yellow coated PF5 flash bulbs fired via an Ilford Junior flash-gun. He later joined ICI Explosives and, in 1956, took underground photographs at Geevor and South Crofty using the Ilford equipment. Unfortunately he did not take any underground photographs at Great Condurrow or at the Holman Test Mine.

George Ellis photographed the Holman winding engine (Plate 59) at Castle-an-Dinas in 1946. Ellis was born in 1900 and became an aircraft engineer before taking up photography. He was a well known press photographer in Middlesex before moving to Cornwall in 1939 to work for the Cornish Guardian, based in Bodmin. He took some underground photographs at South Crofty and Castle-an-Dinas[13]. During the second world war he even took a party of children underground at Redmoor Mine, near Callington, and posed them purportedly at work. This picture was produced as a postcard which the children could then send to their fathers who were fighting abroad. His work was nearly all taken on quarter plate press cameras utilising glass plates. George Ellis died in 1985. His huge collection of negatives are preserved at the Cornish Studies Library in Redruth.

11 Past C.S.M. students will be pleased to note that by contrast the reply and brochure from the Royal School of Mines was "austere and unimpressive".
12 Photograph used on the cover of Cornish Mining by Earl. Revised edition, Cornish Hillside Publications, 1994.
13 See "Castle-an-Dinas 1916-1957, Cornwall's Premier Wolfram Mine" by Tony Brooks, Cornish Hillside Publications, 2001.

John Watton trained in photography at Birmingham College of Photography under the direction of ex-Coal Board photographer George Martin. He went into advertising photography whilst developing an interest in mining, before joining the School of Mines as Head of Audio Visual Aids in 1973. Just as Burrow's research influenced his own lighting technique, so in turn the Burrow and Hughes images influenced Watton's work, already well developed through his work in advertising photography

He was awarded, in 1986, the second only Fellowship of the Royal Photographic Society for a submission of underground Cornish mining scenes. The first such Fellowship was awarded some 91 years earlier to John Charles Burrow.

A number of images (Plates 102-107) were taken underground at Great Condurrow by Watton, during his time at the School. The lighting for these were achieved by a variety of methods including magnesium flash powder with electronic flash to 'fill in' the shadows. The techniques employed in advertising lighting to give subjects shape, depth and form were applicable to underground photography although it required lengthy experimentation to create effective back-lighting. Exposure experimentation was also crucial since reflectivity varied from mine to mine, and indeed from location to location within the same mine.

The camera equipment used by Watton for colour transparencies was a Nikon FTN2 with wide angle lens. For colour or monochrome negatives a large format roll film camera (Mamiya RB67 or Bronica ETRS) or large format sheet film camera. The latter is usually a 5 ins. by 4 ins. Sinar monorail or a 5-ins x 4-ins MPP. Both will fit into a large rifle ammunition can for transportation.

Large flash bulbs are sometimes employed, PFl00, PF60 or M22B where it may not be advisable to fire magnesium. These bulbs are very powerful allowing apertures of f22 or f32 to be used in small areas and about half of those values in larger areas. Where excellence in image quality is paramount the combination of large bulbs, electronic flash fill in and 5 ins. by 4 ins. cameras gives results comparable with those attained by Burrow and Hughes.

Working underground today is much the same as it was at the end of the 19th Century. Camera equipment may be more portable but, having said that, large format technical cameras still have a very important part to play. A 'gang' of helpers is still required to help transport the equipment, fire flash guns or, on rare occasions, magnesium. Often one of the helpers is used as a model since the images are nearly always posed, particularly so when back lighting is employed since it requires a build up of several flashes to achieve the right effect.

Large open areas in old stopes are equally demanding on the light produced and in this situation magnesium is used to fill a large area with light. Of course this technique is restricted to black and white photography, colour work requires the light to be of a certain colour temperature for which the film is balanced to avoid colour casts.

This chapter has only looked at photographers known to have been involved with King Edward Mine. In recent years a number of competent photographers have worked underground recording both the accessible parts of abandoned mines, and the working mines before their closure. Special mention must be made of the superb photographs of Paul Deakin whose work can be found in a number of recent publications.

Appendix III lists the photographers and locations of original prints and negatives of the images used in this book.

What is the future underground photography in Cornwall? The working mines are now all closed and South Crofty may or may not work again. The future seems to be photography in disused mines, but the fascination of taking pictures underground will always be there to lure photographers into the darkness.

Appendix One

List of the capital expenditure, 1897-1902, to bring King Edward Mine to the production stage.

	£	£
Purchase of sett	325	325
Rock drill plant		
Air compressor and receiver	325	
3-ins air-main & 2-ins extensions	60	
House foundations & engine loadings	110	
Steam pipes & misc. accessories	62	
5 Rockdrills & accessories	nil	557
Winding Engine		
Driving wheels etc.	30	
Brake indicator, house over cage[1]	25	55
Compound Engine		
Holman Bros. £635 (less £200)	425	
House + engine loading	150	
Steam pipes & covering, copper expansion joint	50	625
Stamps		
Fraser & Chalmers + shipping & carriage	520	
House & foundations for stamps	200	
Extra shafting & ironwork for frame	25	
Buss table List price £105, for	50	
Frue vanner	nil	795
Surveying		
Drawing office & furniture	400	
Safe	15	
Heating system	25	
Lantern slide & blue-print gear	25	
Surveying instruments	nil	465

1 'Cage' is an old Cornish mining term that was used to describe the winding drum on a hoist.

	£	£
Other Buildings		
General office, vanning office & furniture Carpenter shop, Smith's shop Store, Brass casting shop, Oil brass furnace	260	
Houses over William's & North Shaft	15	275
Dry		
Baths, basins, showers, floor 7 Drains	61	
Structural alterations	35	96
Water Supply & Drainage	75	75
Underground & Shaft		
Re-timbering shaft	100	
Headgear	60	
Skip roads & ladderways (3)	75	
Cages	25	
Water skip	10	
Turn-outs (5)	25	
Keps (6)	12	
Waggons	39	
Rails & sleepers	101	
Air winch, hand winch, blocks etc.	65	512
	Total	£3780

Appendix Two

The Vanning Assay

The sample to be assayed was dried and small representative part of the sample was weighed out on a balance before being concentrated on a vanning shovel. On the vanning shovel the weighed sample was mixed with water and swirled around much as a gold prospector would use a pan. In the hands of a skilled operator the lighter waste would be gradually washed away off the shovel leaving a concentrate of the dense minerals – sulphides of copper, iron and arsenic plus the tin.

The shovel was then placed over the fire to dry the sample which was then brushed off the shovel into a crucible and heated to redness. This roasting, doing the same job as calcination, drove off the sulphur and arsenic thus leaving the light oxides of copper and iron and the unaltered tin. The sample was returned to the vanning shovel, mixed with a little water and ground by rubbing with a small hammer and reconcentrated. After drying the sample was brushed onto a watch glass, stirred with a magnet to remove any iron minerals, and finally weighed on the assay balance.

A comparison between the weight of the ore sample going onto the shovel and the weight of the concentrate from the shovel gave the percentage of tin mineral in the original sample. The shovel vanning mimicked the operation of the shaking tables in the mill and gave a good indication of how much tin would be recovered in the plant.

Appendix Three

Details of the photographs used in this book listing photographer and locations of original prints and negatives where known.

Key
AWB Tony Brooks
BE Bryan Earl
CSL Cornish Studies library – Reduth
CRO Cornwall County Records Office – Truro
CSM Camborne School of Mines
RIC Royal Institution of Cornwall – Truro
WJW John Watton

Plate No.	Photographer	Location of original negative	Location of original or best print	Notes
1	J C Burrow		CSL	
2	W J Bennetts	CSL		neg BT156Q
3	W J Bennetts	CSL		neg BT50Q
4	J C Burrow	RIC	CSM/WJW	RIC neg 02
5	J C Burrow	RIC	CSM/WJW	RIC neg 04
6	J C Burrow	RIC	CSM/WJW	RIC neg 18
7	J C Burrow		CSM/WJW	
8	J C Burrow		CSM/WJW	
9	JCB or HH		CSM/KEM	KEM Museum
10	J C Burrow		AWB	
11	JCB/H Hughes		CSM	1899 Prospectus
12	W Thomas		CSL	Thomas album
13	J C Burrow	RIC	CSM/WJW	RIC neg 07
14	J C Burrow	RIC	CSM/WJW	RIC neg 09
15	J C Burrow		WJW	
16	H Hughes	RIC	CSM/WJW	RIC neg 24
17	JCB or HH		CSM	1899 Prospectus
18	J C Burrow		WJW	
19	J C Burrow	RIC	CSM/WJW	RIC neg 19
20	H Hughes	RIC	CSM/WJW	RIC negs 01 & 14
21+	J C Burrow		CSM/CRO	lantern slide CRO ref. AD460/1/4
22	H Hughes	RIC	CSM/WJW	RIC neg 15

+ Print made from CSM lantern plate not from CRO plate found recently

Plate No.	Photographer	Location of original negative	Location of original or best print	Notes
23	J C Burrow		CSM/WJW	
24	J C Burrow	RIC	CSM/WJW	RIC neg 05
25	J C Burrow	RIC	CSM/WJW	RIC neg - listed unidentified
26	H Hughes	RIC	CSM/WJW	RIC neg 28
27	J C Burrow		CSM*	lantern slide
28	H Hughes	RIC	CSM/WJW	RIC neg 21
29	J C Burrow		CSM*	lantern slide
30	H Hughes	RIC	CSM/WJW	RIC neg 22
31	J C Burrow		CSM/WJW	
32	N Ward		CSL	copy neg BT 129H
33	H Hughes	RIC	CSMWJW	RIC neg 13
34	?		CSL	Thomas album
35	J C Burrow		CRO	CRO AD 460/12
36	J C Burrow		CSL	Thomas album
37	W J Bennetts	CSL	WJW	neg BT 46Q
38	J C Burrow		WJW	
39	J C Burrow		CSM/KEM	KEM Museum
40	W J Watton	CSM*	CSM/WJW	
41	W J Watton	CSM*	CSM/WJW	
42	W Thomas?		CSL	Thomas album
43	W Thomas?		CSL	Thomas album
44	J C Burrow	CSM*	WJW	Lantern slide
45	H A Burrow		CSL	Thomas album
46	JCB/HH?		CSM/KEM	KEM Museum
47	J W Chenhall		AWB	Trans CIE 1918
48	H Hughes	RIC	CSM/WJW	RIC neg 12
49	J C Burrow	CSL	CSM/KEM	copy neg BT 82H
50	J C Burrow		CSM/KEM	KEM Museum
51	W J Bennetts		AWB	Post card
52	H Hughes		CSM	Lantern slide
53	W J Bennetts	CSL	CSM	neg BT 87H
54	?		AWB	Holman catalogue 1908
55	?		CSM	CSM Journal
56	?		E Sharpley	AWB copy
57	H Hughes	RIC	CSM/WJW	RIC neg 31
58	H Hughes	RIC	CSM/WJW	RIC neg 11
59	G Ellis	CSL	AWB	neg E10739
60	W J Bennetts	CSL	WJW	neg BT 23Q
61	W J Bennetts		P Bradley	post card
62	?		AWB	62-68 taken by

Plate No.	Photographer	Location of original negative	Location of original or best print	Notes
63	Not known		AWB	an unnamed student
64	Not known		AWB	(possibly R H Tredgold)
65	Not known		AWB	
66	Not known		AWB	
67	Not known		AWB	
68	Not known		AWB	
69	Not known		CSM	
70	A Jukes-Brown?	J Rapson	AWB	
71	Not known	CSM*	CSM	
72	F Hutchin	AWB	AWB	
73	Assoc. Press		AWB	
74	Assoc. Press		AWB	
75	Mine & Quarry Eng.	AWB		
76	W J Bennetts & Sons	CSL	WJW	neg BT 88H
77	W J Bennetts & Sons		CSM	Prospectus. 76-78 taken
78	W J Bennetts & Sons		BE	by W J Bennetts' son.
79-83	Not known		CSM	79-83 in CSM 1937 prospectus probably by Bennetts & Son
84	Mine & Quarry Eng.	AWB		
85	G J Nicholas?		CSM	1954 prospectus
86	Not known		AWB	
87	Not known		CSM	1946 prospectus
88	Not known		CSM	1946 prospectus
89	Not known		AWB	
90	G J Nicholas?		CSM	1954 prospectus
91	B Earl	BE	BE/AWB/WJW	
92	B Earl	BE	BE/AWB/WJW	
93	B Earl	BE	BE/AWB/WJW	
94	B Earl	BE	BE/AWB/WJW	
95	C V Smale	CVS	AWB	
96	A J Clarke	CSM	CSM/KEM	KEM Museum
97	Not known		AWB	
98	C V Smale	CVS	AWB	
99	W J Watton	CSM*	CSM/WJW	
100	W J Watton	CSM*	CSM/WJW	
101	W J Watton	CSM*	CSM/WJW	
102	W J Watton	CSM*	CSM/WJW	
103	W J Watton	CSM*	CSM/WJW	
104	W J Watton	CSM*	CSM/WJW	

* Negatives now missing

Plate No.	Photographer	Location of original negative	Location of original or best print	Notes
105	W J Watton	CSM*	CSM/WJW	
106	W J Watton	CSM*	CSM/WJW	
107	W J Watton	CSM*	CSM/WJW	
108	A W Brooks	AWB	AWB	
109	A W Brooks	AWB	AWB	
110	A W Brooks	AWB	AWB	
111	L Ankers	CSM	AWB	
112	A W Brooks	AWB	AWB	
113	A W Brooks	AWB	AWB	
114	A W Brooks	AWB	AWB	
115	A W Brooks	AWB	AWB	
116	A W Brooks	AWB	AWB	
117	A W Brooks	AWB	AWB	
118	A W Brooks	AWB	AWB	
119	A W Brooks	AWB	AWB	
120	A W Brooks	AWB	AWB	
121	A W Brooks	AWB	AWB	
122	W J Bennetts		AWB	Postcard
123	A W Brooks	AWB	AWB	
124	A W Brooks	AWB	AWB	
125	L Ankers	CSM	AWB	
126	J C Burrow		CRO	CRO ref AD 460/1/6
127	T Roskrow		P Bradley	Postcard

* negatives now missing

Index